The Political Economy of Low Carbon Resilient Development

Over the last decade, policies and financing decisions aiming to support low carbon resilient development within the least developed countries have been implemented across several regions. Some governments are steered by international frameworks, such as the United Nations Framework Convention on Climate Change (UNFCCC), while others take their own approach to planning and implementing climate resilient actions. Within these diverse approaches, however, there are unspoken assumptions and normative assessments of what the solutions to climate change are, who the most appropriate actors are and who should benefit from these actions.

This book examines the political economy dynamics or the underlying values, knowledge, discourses, resources and power relationships behind decisions that support low carbon resilient development in the least developed countries. While much has been written on the politics of climate change, this book focuses on the political economy of national planning and the ways in which the least developed countries are moving from climate resilient planning to implementation. The book uses empirical evidence of low carbon resilient development planning in four countries: Bangladesh, Ethiopia, Rwanda and Nepal. Different approaches to low carbon resilience are critically analysed based on detailed analysis of key policy areas.

This book will be of great interest to policy makers, practitioners' students and scholars of climate change and sustainable development.

Neha Rai is a Senior Researcher with the Climate Change Group at the International Institute for Environment and Development (IIED).

Susannah Fisher is a Senior Researcher with the Climate Change Group at the International Institute for Environment and Development (IIED).

Routledge Studies in Low Carbon Development

Low Carbon Transitions for Developing Countries
Frauke Urban

Towards Low Carbon Cities in China
Urban form and greenhouse gas emissions
Sun Sheng Han, Ray Green and Mark Wang

The Political Economy of Low Carbon Transformation
Breaking the habits of capitalism
Harold Wilhite

The Social Challenges and Opportunities of Low Carbon Development
Johan Nordensvärd

The Political Economy of Low Carbon Resilient Development
Planning and implementation
Edited by Neha Rai and Susannah Fisher

The Political Economy of Low Carbon Resilient Development
Planning and implementation

Edited by
Neha Rai and Susannah Fisher

First published 2017
by Routledge
2 Park Square, Milton Park, Abingdon, Oxon OX14 4RN

and by Routledge
711 Third Avenue, New York, NY 10017

Routledge is an imprint of the Taylor & Francis Group, an informa business

© 2017 selection and editorial matter, Neha Rai and Susannah Fisher; individual chapters, the contributors

The right of the editors to be identified as the authors of the editorial material, and of the authors for their individual chapters, has been asserted in accordance with sections 77 and 78 of the Copyright, Designs and Patents Act 1988.

All rights reserved. No part of this book may be reprinted or reproduced or utilised in any form or by any electronic, mechanical, or other means, now known or hereafter invented, including photocopying and recording, or in any information storage or retrieval system, without permission in writing from the publishers.

Trademark notice: Product or corporate names may be trademarks or registered trademarks, and are used only for identification and explanation without intent to infringe.

British Library Cataloguing in Publication Data
A catalogue record for this book is available from the British Library

Library of Congress Cataloging in Publication Data
Names: Rai, Neha, editor. | Fisher, Susannah, 1983- editor.
Title: The political economy of low carbon resilient development planning and implementation / edited by Neha Rai and Susannah Fisher.
Description: Abingdon, Oxon ; New York, NY : Routledge, [2017] | Series: Routledge studies in low carbon development
Identifiers: LCCN 2016015666 | ISBN 9781138932975 (hb) | ISBN 9781315679280 (ebook)
Subjects: LCSH: Carbon dioxide mitigation--Economic aspects--Developing countries. | Economic development--Environmental aspects--Developing countries. | Environmental policy--Economic aspects--Developing countries. | Climatic changes--Economic aspects--Developing countries.
Classification: LCC HC59.72.E5 P65 2017 | DDC 338.9/27091724--dc23
LC record available at https://lccn.loc.gov/2016015666

ISBN: 978-1-138-93297-5 (hbk)
ISBN: 978-1-315-67928-0 (ebk)

Typeset in Goudy
by Taylor & Francis Books

Contents

	List of illustrations	vii
	Acknowledgements	ix
	Notes on contributors	x
	Abbreviations	xiii
1	Understanding the politics of low carbon resilient development in the least developed countries NEHA RAI AND SUSANNAH FISHER	1
2	National and international experiences of low carbon resilient development SUSANNAH FISHER, DAVE STEINBACH, COSTANZA POGGI AND SALEEMUL HUQ	22
3	Storylines in national climate planning and politics: finance, ownership and shifting synergies SUSANNAH FISHER, MISGANA E. KALLORE, NAZRIA ISLAM, LIDYA TESFAYE AND JOHN RWIRAHIRA	47
4	National political economy of climate funds: case studies of the PPCR and the SREP NEHA RAI AND THOMAS TANNER	65
5	Designing climate finance systems in Ethiopia and Rwanda NANKI KAUR, DANIEL FIKEREYSUS, JOHN RWIRAHIRA, LIDYA TESFAYE AND SIMRET MAMUYE	88
6	Incentives and opportunities for local energy finance NEHA RAI AND DAVE STEINBACH	106
7	Using political economy analysis as a tool in national planning NEHA RAI AND ERIN J. NASH	131

8 Supporting effective low carbon resilient development: Lessons from the political economy of the least developed countries 152
SUSANNAH FISHER

Index 171

Illustrations

Figures

1.1	Analytical framework used in this book	13
4.1	UNFCCC and non-UNFCCC multilateral climate funding	68
4.2	Total PPCR funding and co-financing in 2014, by investment type	71
4.3	SREP total funding by country and technology type at 2013	79
5.1	Policy options for mobilising and delivering finance for inclusive investment in LCRD	99
5.2	Coalitions supporting approaches to financing inclusive investment in LCRD	101
7.1	DFID's Drivers of Change approach	133
7.2	Typical steps of a political economy analysis	137
7.3	A power-interest grid for stakeholder mapping	142
7.4	Results of a stakeholder analysis of the PPCR initiative in Bangladesh	143
7.5	Actors' narratives about how PPCR will bring about 'transformational change' in Bangladesh	146
7.6	Climate finance landscape framework	148
7.7	General procedure for applying the findings of a political economy analysis	149

Tables

2.1	Main UNFCCC planning frameworks affecting LDCs	24
3.1	Flagship LCRD plans in Bangladesh, Ethiopia and Rwanda	50
3.2	Storylines supporting LCRD in Bangladesh, Ethiopia and Rwanda	51
3.3	Understandings of LCRD in Ethiopia, Bangladesh and Rwanda: storylines and coalitions	54
4.1	PPCR investments in Nepal and Bangladesh	73
4.2	Narratives and incentives related to PPCR investment decisions in Bangladesh and Nepal	75

4.3	SREP investment portfolio, Ethiopia and Nepal	80
4.4	Narratives and incentives related to SREP investment decisions in Ethiopia and Nepal	82
5.1	Global flows of climate finance	93
6.1	Public-private investment in decentralised renewable energy in four LDCs	110
6.2	Intermediaries engaged to deliver LCRD and their characteristics	116
6.3	Incentives for actors to invest in renewable energy	120
6.4	Examples of extent of effectiveness of pro-poor measures from renewable energy projects in Bangladesh, Ethiopia, Nepal and Rwanda	123
7.1	Institutions and governance arrangements for implementing the PPCR in Bangladesh and Nepal	139
7.2	Categories of stakeholders engaged in the design and implementation of the PPCR in Bangladesh and Nepal	142
7.3	Summary of the main incentives driving PPCR priorities in Bangladesh	147

Boxes

2.1	Main features of climate change planning in Bangladesh	32
2.2	Main features of climate change planning in Ethiopia	35
2.3	Main features of climate change planning in Rwanda	38
2.4	Main features of climate change planning in Nepal	40
5.1	Ethiopia's Coffee Initiative	99
6.2	Special purpose agencies promoting renewable energy provision in low-income off-grid communities	112
6.2	Central and national development bank	113

Acknowledgements

We have worked together as equal co-editors to bring together the findings of this collaborative research project and we would like to extend our personal gratitude to all those who helped us shape this book.

Firstly, we would like to thank the government agencies in Bangladesh, Ethiopia, Nepal and Rwanda who supported our work through interviews, attending workshops and informal discussions as well as all our interviewees who gave their time and thoughts. Thanks also to our reviewers Tim Forsyth and Gareth Edwards for their useful comments which greatly strengthened the analysis. Any errors and omissions remain our own.

We would like to thank DFID and UK Aid for the financial support for the IIED programme of work on understanding the political economy of climate resilient development which significantly contributed to the empirical basis of this book. We would also like to thank IIED for giving us the opportunity to publish this book, and to IIED's country partners who worked closely with us on all aspects of the research: Clean Energy Nepal, Echnoserve, IPAR Rwanda, Bangladesh Centre for Advanced Studies, the Centre for Participatory Research and Development and the International Centre for Climate Change and Development.

Finally and above all, we would like to extend our sincere thanks to all members of IIED's Public Policy Team in the Climate Change Group who supported this research and the publication of this book at different stages including Hohit Gebreegziabher, Simon Anderson, Nanki Kaur and Dave Steinbach.

Susannah Fisher and Neha Rai

Contributors

Sunil Acharya is an environment and sustainable development policy analyst based in Kathmandu, Nepal. He is co-founder and director of Digo Bikas Institute, a research and advocacy organisation committed to promoting ecological sustainability and social equity at policy and community level.

Raju Pandit Chhetri is associated with Prakriti Resources Centre, an NGO based in Kathmandu, Nepal, that promotes climate change policy and environment-friendly development practices. Raju also works to strengthen national and international climate change policy processes, with a particular focus on climate finance, local adaptation and low carbon initiatives in Nepal. He also keenly follows the progress of the Green Climate Fund.

Ramesh Bhushal is an environment journalist and researcher based in Nepal. Over the course of the past decade he has contributed to many national and international media outlets. As an insider witness of UN climate change negotiations since 2009, he also served as a negotiator for the Nepalese government at the UNFCCC.

Susannah Fisher is a senior researcher at the International Institute for Environment and Development (IIED), where she works on issues of climate change politics and governance in the global South with a focus on climate justice, adaptation effectiveness and gender equality. She has a PhD from Cambridge University and has also worked in the Grantham Institute at the London School of Economics and the United Nations Economic Commission for Africa.

Daniel Fikereysus is a senior researcher and CEO of Echnoserve Consulting, an Ethiopia-based sustainable development, environment and energy consulting firm, with over seven years of experience working on climate change issues.

Saleemul Huq is the director of the International Centre for Climate Change and Development based at the Independent University, Bangladesh, and is also senior fellow at the London-based IIED. He is a global expert on climate resilient development pathways and has worked extensively with and in the least developed countries of Africa and Asia, particularly on community-based adaptation to climate change.

Nazria Islam is a specialist on adaptation and climate change in Bangladesh. She was previously Senior Research Officer at the Bangladesh Centre for Advanced Studies, and now works at the Bangladesh Rural Advancement Committee as Senior Manager, Climate Change and Livelihood, within the Disaster Management and Climate Change Programme.

Misgana E. Kallore is a co-author on several research publications with IIED and others. She is based in Ethiopia and previously worked with Echnoserve Consulting. In 2015 she joined the Deutsche Gesellschaft für internationale Zusammenarbeit (GIZ) as a national coordinator of Green Economy Transition in Africa, a joint GIZ/UNEP project. She works to support district-level planning unit experts to integrate and deliver the green economy agenda. She works closely with Ethiopia's Ministry of Environment, Forest and Climate Change.

Nanki Kaur is a principal researcher with the Climate Change Group at the IIED. She leads the team on public policy responses to climate change. She has over 14 years of experience in the areas of pro-poor development, climate change planning, natural resource governance and aid effectiveness in South Asia and East Africa.

Simret Mamuye is a researcher working on climate change and sustainable development at Echnoserve Consulting. She is currently engaged in the IIED project on the political economy of climate finance. She has over six years of experience in working with public organisations on public finance, climate change, planning, monitoring and evaluation.

Erin J. Nash is a consultant researcher for IIED's Climate Change Group, and a PhD candidate at the Centre for Humanities Engaging Science and Society in the Department of Philosophy, Durham University, UK. Erin has more than 12 years of experience as a researcher, policymaker, and practitioner, and has previously worked for the Australian government and for the Worldwide Fund for Nature in Southeast Asia. She has a BSc from Monash University and an MSc (Philosophy and Public Policy) from the London School of Economics and Political Science.

Costanza Poggi is a policy assistant at the London-based think tank Green Alliance. Her focus is the political leadership theme, which aims to increase political action and dialogue on climate change in the UK by working with political and business leaders as well as environmental NGOs. She holds a BA in International Relations and an MA in Environment, Development and Policy, both from the University of Sussex, UK.

Neha Rai is a senior researcher with the Climate Change Group at the IIED. Based in the UK, she leads the IIED's work on the political economy of financing low carbon resilient development. Her areas of technical expertise include in-country work on the political economy dynamics of international financing, decentralised energy access and local-level financing.

John Rwirahira is a senior researcher at the Institute of Policy Analysis and Research in Rwanda (IPAR-Rwanda), working on issues of public policy, local governance, climate and natural resources.

Md Shamsudoha is the chief executive of the Centre for Participatory Research and Development, a research-based NGO in Bangladesh. He is a member of the Bangladesh delegation to the UNFCCC climate change negotiations and has particular interest in negotiations on loss and damage, finance, adaptation and technology transfer. He is involved in the implementation of a number of projects with different academic institutions, in-country and abroad, focusing in particular on climate change, disaster risk reduction and coastal livelihoods.

Dave Steinbach is a researcher with the Climate Change Group at the IIED. He works on public policy responses to climate change in South Asia and Africa, focusing on social protection and climate resilience, financing low carbon resilient development and livelihood transitions for rural people working in agriculture and forestry and fisheries. He has an MSc in International Development Studies from the London School of Economics and Political Science.

Thomas Tanner is a team leader for the Climate and Environment Programme at the Overseas Development Institute, where he leads work on adaptation and resilience. He is a development geographer working on adaptation and resilience to the impacts of climate change, including the politics of climate change policy processes; organisational change; poverty, vulnerability and climate adaptation; children and disasters; and building resilience in urban contexts. He has 20 years of experience as a researcher, policy-maker, practitioner and negotiator at UN conventions on climate and desertification.

Lidya Tesfaye is a researcher working on climate change and sustainable development at Echnoserve Consulting. Based in Ethiopia, she has been involved in an IIED study of the political economy of climate resilient development, coordinating with key stakeholders and conducting data collection and analysis. She also has a good understanding of international and national climate finance issues.

Abbreviations

AEPC	Alternative Energy Promotion Centre (Nepal)
BCCRF	Bangladesh Climate Change Resilience Fund
BCCSAP	Bangladesh Climate Change Strategic Action Plan
BCCTF	Bangladesh Climate Change Trust Fund
CCIOU	Climate Change and International Obligations Unit
CIF	Climate Investment Fund
COP	Conference of the Parties (UNFCCC)
CREF	Central Renewable Energy Fund (Nepal)
CRGE	Climate Resilient Green Economy (Ethiopia)
DFID	Department for International Development (UK)
EDPRS	Economic and Development Poverty Reduction Strategy (Rwanda)
FONERWA	Fonds National de l'Environnement (Rwanda's national fund for environment and climate change)
GCF	Green Climate Fund
GDP	gross domestic product
GGGI	Global Green Growth Institute
GTP	Growth and Transformation Plan (Ethiopia)
IDCOL	Infrastructure Development Company Ltd (Bangladesh)
IDS	Institute of Development Studies
IFC	International Finance Corporation
IIED	International Institute for Environment and Development
INDC	Intended Nationally Determined Contributions
IPCC	Intergovernmental Panel on Climate Change
LAPA	Local Adaptation Plan of Action
LCRD	low carbon resilient development
LDCs	least developed countries
LDCF	Least Developed Countries Fund
MINECOFIN	Ministry of Finance and Economic Planning (Rwanda)
MINIRENA	Ministry of Natural Resources (Rwanda)
MOEF	Ministry of Environment and Forest (Bangladesh)
MOFED	Ministry of Finance and Economic Development (Nepal)

MOSTE	Ministry of Science, Technology and Environment (Nepal)
MOWIE	Ministry of Water, Irrigation and Energy (Nepal)
NAMA	Nationally Appropriate Mitigation Action
NAP	National Adaptation Plan
NAPA	National Adaptation Programme of Action
NGO	non-governmental organisation
NRREP	National Rural Renewable Energy Programme (Nepal)
NSCCLCD	National Strategy for Climate Change and Low Carbon Development (Rwanda)
ODA	overseas development assistance
OECD	Organisation for Economic Co-operation and Development
ODI	Overseas Development Institute
PPCR	Pilot Program for Climate Resilience
REMA	Rwanda Environment Management Authority
SCCF	Special Climate Change Fund
SIDA	Swedish International Development Agency
SME	small and medium-sized enterprises
SPCR	Strategic Program for Climate Resilience
SREP	Scaling Up Renewable Energy Program
UK	United Kingdom
UN	United Nations
UNDP	United Nations Development Programme
UNEP	United Nations Environment Programme
UNFCCC	United Nations Framework Convention on Climate Change
UN-REDD	United Nations collaborative initiative on Reducing Emissions from Deforestation and forest Degradation

1 Understanding the politics of low carbon resilient development in the least developed countries

Neha Rai and Susannah Fisher

Introduction

Climate change caused by greenhouse gas emissions is affecting societies and ecosystems across the globe. For decades scientists have expressed grave concerns about the implications of global warming. The Intergovernmental Panel on Climate Change (IPCC) estimates that a rise in average temperature of 2°C above the pre-industrial value (approximately 280 ppm) will significantly affect sea levels, food supply, health and extreme weather events (IPCC 2014). Although the issue is now mainstreamed within policy debates, it has remained contested for many years. Despite increasing evidence and urgency, advocates have struggled to get policymakers, citizens and even some scientists to take this threat seriously (Giddens 2011). A small number of scientists believe that climate change is not the result of human activity but rather of natural shifts in the environment (Curry and Webster 2011; Dyck et al. 2007). Other researchers believe that social issues such as health, education and poverty are more urgent (Goklany 2005, 2009). There was also an initial lack of urgency among policymakers and the wider public over pursuing climate change responses; Giddens (2011) has described this as a paradox, with a lack of tangible evidence inhibiting further action. As demonstrated by decades of climate change debate, knowledge and evidence are repeatedly used to steer or amplify agendas according to the interests of those involved.

Policymakers around the world are now increasingly concerned about the scale and urgency of climate change, and there is broad agreement within scientific communities regarding its likely impact. As a result many countries are developing plans to deal with this global threat. The growing consensus about the existence of the problem does not imply that climate change has become any less political, however. Climate change debates and actions remain highly complex, with many actors, agencies and institutions steering in different directions. Current political challenges relating to climate change centre around the north–south divide in negotiations, with 'Annex 1' countries (industrialised Organisation for Economic Co-operation and Development (OECD) countries according to the United Nations Framework Convention on Climate Change (UNFCCC)) demanding that emerging economies such as India and the People's Republic of China which have larger shares in global emissions take a similar level of responsibility for climate action. Meanwhile, developing countries maintain that those countries historically responsible for causing the damage should finance their mitigation and adaptation efforts. Recognising the changing nature of the debate, this book shows how understanding the political economy of climate change is crucial to addressing the issue successfully.

Much of the existing research, including that dealing with developing countries, focuses only on apolitical and technocratic aspects of addressing climate change. With some notable exceptions (discussed later in this chapter), studies have targeted scientific, often linear solutions, and have paid limited attention to the political nature of the national governance systems needed to implement these solutions. A political economy approach to understanding climate change responses can provide much deeper insights into the efficiency, equity and effectiveness of the processes involved, and has particular relevance for low-income and developing countries. It is clear that climate change will affect some countries more than others: the least developed countries (LDCs), which have the lowest levels of socioeconomic development globally, will experience more negative social, economic and environmental impacts and have fewer resources for dealing with them. In addition, LDCs are often poorly placed to engage with the political and administrative processes involved in accessing finance or implementing technical solutions. On top of this, they must contend with ever changing international agendas and the difficulty of making their voices heard.

It is in this evolving context that countries have started identifying ways to tackle climate change, using what is referred to as low carbon resilient development (LCRD). This describes an approach to development that seeks to integrate climate change mitigation (reducing greenhouse gas emissions), climate change adaptation (reducing the adverse effects of climate change) and sustainable development. The LCRD agenda was recognised in the fourth IPCC assessment report (IPCC 2007a) and gained further momentum when countries began developing their own LCRD frameworks. At both national and international level, and particularly in LDCs, policymakers are now actively devising strategies, plans and programmes that combine climate resilience with reduced carbon emissions.

This transition has been influenced by several factors. First, LDC governments are increasingly aware of – and indeed experiencing – the impact of climate change; this includes its impact on development, such as growing poverty, loss of livelihoods and reduced socioeconomic well-being. Second, governments are also keen to harness the available international climate finance to achieve co-benefits from LCRD (Boyle 2013, Fisher et al. 2014). Third, many funders, particularly donor countries, have emphasised the integration of the mitigation and adaptation agendas as a means of ensuring the efficiency and coherence of climate change responses; treating these agendas as separate is seen as a false dichotomy (Dessler and Parson 2010).

Following the international climate change agreement reached in Paris, France, in 2015, the focus of action is shifting from international debates to national and local governments. These governments now need to deliver on the ambitious promises made to reduce emissions and secure a climate resilient future for their populations. Understanding how these promises are likely to translate into action at national level is crucial, as is identifying the actions which have the best chance of success.

In this book we analyse how national governments in LDCs have started to identify LCRD solutions to climate change, including the financing and planning arrangements they put in place. Rather than taking a 'surface' approach, i.e. one that itemises institutions, policies and associated outcomes, we use a political economy lens to examine the actors, ideas, values and incentives that underlie these policy responses. Political economy drivers are the international- and national-level factors that shape the success or failure of climate change strategies and frameworks (Tanner and Allouche 2011). Actors' interests, knowledge, networks and incentives help to determine whether LCRD responses are inclusive and reach the low-income communities that need most support. We argue that to ensure effective and equitable outcomes, actors must understand the influence of the political economy at multiple levels of analysis and governance.

While much has been written on the international politics of climate change (see Adger 2003; Tanner and Mitchell 2008; Giddens 2011), few studies examine how mitigation, adaptation and sustainable development agendas are brought together in national policies. Countries are using a variety of approaches to LCRD planning, including national strategies, Intended Nationally Determined Contributions[1] (INDCs), development planning frameworks, green growth plans and climate change action plans. At the same time, institutional responsibility for climate change within countries has shifted. Whereas previously environment ministries were solely responsible for climate change responses, increasingly this responsibility is being distributed across agencies. A variety of special implementation arrangements have also been developed, including dedicated financial agencies and programmes such as Nationally Appropriate Mitigation Actions (NAMAs).

While it is possible to identify the emergence of various distinct approaches to LCRD, the process is neither uniform nor straightforward. As countries move from planning to implementation there are discrepancies between the visions they set out to achieve and reality. In some cases the adaptation and mitigation agendas continue to operate within separate policy 'silos', due to the practical and financial difficulty of linking them together. National governments taking a cross-sectoral approach also struggle to resolve priorities across the various ministries and departments involved.

By viewing national planning and implementation processes from a political economy perspective, this book provides new insights into the politics of

LCRD and climate change. In the following sections we explore the way in which LCRD is framed and review the existing literature on the topic; we also introduce the analytical framework and methods used in this book and provide an overview of its structure and the main findings. Chapter 2 then looks at the broad trends from national and international experiences on LCRD, before subsequent chapters present case studies of LCRD experiences in four of the LDCs.

What is low carbon resilient development?

As mentioned earlier, LCRD is a policy paradigm designed to enable developing countries to integrate their climate change mitigation and adaptation responses and their development actions. The interplay between mitigation, adaptation and development was acknowledged in the IPCC's *Fourth Assessment Report* (IPCC 2007a) and the benefits of synergising these separate strands have been recognised subsequently. A mitigation project, for example, is likely to be more sustainable if it also builds the climate change resilience of the communities involved; this can also increase local 'ownership' of the project. Similarly, an adaptation project contributing to emission reductions may be able to access greater funding and build capacities by drawing on more diverse resources.

To describe the three elements of LCRD in more detail:

- Mitigation involves actions to reduce anthropogenic (human-induced) emissions of greenhouse gases, which include water vapour, carbon dioxide, methane and nitrous oxide (IPCC 2007b). The five main ways of achieving mitigation are energy efficiency; use of renewable energy; carbon sequestration in sinks such as forests; land management for emission reductions; and geo-engineering such as carbon capture and storage (Boyd and Tompkins 2010). Mitigation opportunities in developed countries are mainly in the energy and transport sectors, while developing countries will make more reductions in the forestry, energy and agriculture sectors (Klein et al. 2007).
- Adaptation involves the 'adjustment of natural systems in response to actual or expected climatic stimuli or their effect' (IPCC 2007a). Adaptation activities fall on a spectrum that ranges from addressing the drivers of vulnerability, to building response capacity, to managing climate risk, to confronting climate change (McGray et al. 2007). Adaptation frequently involves national actions to 'climate-proof' infrastructure or development sectors, as well as local coping strategies for dealing with current variations in climate.
- Development priorities traditionally include outcomes such as improving health and education, and reducing absolute poverty. These aims are closely related to the underlying drivers of vulnerability to climate change: reliance on climate-sensitive sectors such as agriculture and a

lack of capacity for adaptation are both factors that can be addressed through economic growth and development. Some scholars suggest that only the right kinds of growth policies – those that address considerations such as natural capital, institutional and regulatory frameworks, infrastructure, human capital and access to markets – will reduce vulnerability (Bowen et al. 2012). Others reiterate that not all development is also adaptation, nor is all adaptation necessarily development (Ayers and Dodman 2010). While adaptation (particularly that at the 'vulnerability' end of the spectrum) and development are closely linked, the relationship between them is complex. This underscores the need to consider their complementary aims when formulating policy.

In practice, countries are taking different policy approaches to integrating mitigation, adaptation and economic development agendas. Some are establishing broad LCRD strategies while others are weighting their response to one agenda more than another, depending on national priorities. The different ways in which climate change responses are framed in different countries illustrates that the process remains contested. In this book we adopt a broad interpretation of LCRD in order to capture how these contested agendas play out in different contexts. Chapter 4, for example, analyses national experiences of mitigation and adaptation funds, while Chapter 6 examines how different countries are financing decentralised low carbon energy (an example of an LCRD project). Chapter 3 also deals specifically with how synergies between the adaptation, mitigation and development agendas can be found, exploring a co-benefits approach to LCRD.

Why does the political economy of LCRD matter?

The political economy approach used in this book seeks to understand the domestic politics of LCRD within LDCs, which historically are the most susceptible to the social injustices of climate governance. The need for further analysis of the political economy dimensions of LCRD in LDCs relates to four main issues:

- Climate change policymaking has been studied extensively at international level, highlighting issues related to international negotiations. Discussions at national level have also received some attention, but few studies have examined the process of climate planning in low-income countries. Now that countries are beginning to harness global and domestic funds for the purposes of investment in LCRD there is a clear need to understand how domestic politics shapes policy and practice.
- Given that LCRD planning is still relatively new and that achieving synergies between the adaptation and mitigation agendas is not straightforward, a political economy analysis can help us to understand the

benefits and challenges involved, and how decisions can be steered effectively.
- The landscape of international climate finance is changing, with resources increasing and new opportunities and barriers being created. However, the political dynamics involved – which determine whether measures are equitable and inclusive for LDCs – remain relatively underexplored. Understanding the politics of global climate finance and its implications for the LDCs can help to bridge this knowledge gap.
- Funders have for some time been using political economy analysis to evaluate specific projects or programmes, and how countries are (or will be) implementing them. We argue, however, that a more constructive approach is needed, in which political economy analysis is used by countries themselves to steer better, consensus-driven decision making in LCRD processes.

Regarding the treatment of the first of these issues in the existing literature, for decades the analysis of climate change politics has centred on international issues, particularly north–south divisions and questions of power and inclusion in climate change negotiations (see, for example, Paterson and Grubb 1992; Luterbacher and Sprinz 2001; Salehyan and Hendrix 2010). A range of studies raises important issues concerning the responsibilities of polluters, compensation of developing countries and the global costs and benefits of curbing emissions (see, for example, Aldy et al. 2003; Ciplet et al. 2013; Khan 2014; Barrett 2014). Arguments here centre on justice, and claim that those most responsible for climate change should bear its financial and political burdens (Giddens 2011). Paavola and Adger (2006), for example, have argued for justice-based approaches to global climate change politics that recognise the need to 'avoid dangerous climate change', take 'forward-looking responsibility', put the 'most vulnerable first' and ensure the 'equal participation' of all. Although studies like these have provided insight into issues of power, inclusion and equity in international climate change relations, they do not fully address questions about how agendas are influenced by or within the domestic context of LDCs.

While climate change responses at national level have been the subject of research and discussion, the emphasis has been primarily on technological solutions (Tanner and Allouche 2011), perhaps because scientists are at the centre of climate change debate. These studies also tend to assume that governance systems are uniform replicable models and policy processes are consistent and linear (Klein et al. 2007; Leach et al. 2010). A political economy approach can complement existing research by helping us better to understand how and why particular solutions are promoted or adopted in different contexts.

Existing research in this area that examines the way in which national political circumstances encourage policymakers to develop or adopt particular climate strategies is mostly focused on high-income economies, however (see, for example, McCright and Dunlap 2011; Bailey et al. 2012; Hochstetler and Viola 2012). A few scholars have looked at the political economy of climate change in developing countries (for example Ayers et al. 2011; Newell et al. 2011) but most of this work explores the influence of international programmes at national level, rather than exploring the role of domestic political economies in shaping climate change agendas or actions (Dodman and Mitlin 2015 is one exception). Further research is needed, then, into the role of national and local politics in climate change responses in LDCs, and in particular how the mitigation, adaptation and development agendas can be brought together in these contexts.

The increasing scale of international climate finance is another factor here, emphasising the need to understand how low-income countries access, manage and use available resources. The Cancun Agreements, made at the UNFCCC Conference of the Parties (COP16), in November 2010, sought commitments amounting to approximately US $100 billion per year in 'new and additional' climate finance for developing countries. This led to the formation of the Green Climate Fund and the pledging of $30 billion by 2012 in fast-start finance. As a result of these global efforts, a wide range of funds, including the Climate Investment Funds (CIFs) and the Green Climate Fund (GCF), are channelling finance to developing countries (Nakhooda and Norman 2014). These changes are in turn influencing politics in these countries, with new incentives and structures shaping priorities and power relations (Tanner and Allouche 2011). Within countries various actors negotiate for, and influence the delivery of, climate finance; domestic politics are therefore a key determinant of whether and how international decisions are translated into national actions.

In addition, countries are no longer relying solely on international climate finance. Domestic governments are also investing their own money in a wide range of LCRD projects and programmes, using public funds to leverage co-finance from donors and investors. Their internal climate policy planning and processes determine how this money is used, as well as the climate actions taken, how they are implemented and by whom. Domestic politics also constrain or enable countries' involvement in international negotiations (Sprinz and Weiss 2001). From this perspective, there is an urgent need to look beyond international politics and to understand the internal political and administrative processes that influence LCRD decisions, particularly for delivering complex low carbon, climate resilient actions that require joint action.

Turning to the second issue we identified earlier, LCRD is a new, fast-evolving and challenging area of public policy. It involves actors from different policy communities, with intersecting responsibilities and mandates, and requires input from a variety of experts and sources of scientific knowledge. Its cross-sectoral nature creates coordination issues and may require decisions to be made about trade-offs. Arriving at such decisions and setting priorities can be a contested and highly political process. In addition, the high-profile nature of the issues and the amount of resources involved can create tensions and conflicts. Differences in the knowledge levels of interdisciplinary departments and sectors may also affect the extent to which stakeholders can participate in planning processes (Boyle 2013).

As well as these cross-sectoral challenges, LCRD also poses the difficult problem of harmonising adaptation, mitigation and development agendas. In many countries one of these agendas is prioritised over the others (Fisher 2013), according to the country context, resource requirements, growth strategies, and areas and levels of pre-existing expertise. It is rare to find instances of balanced allocation of resources between adaptation and mitigation, and where these exist they are limited to specific sectors such as energy or agriculture (Boyle 2013; Illman et al. 2012). Internal politics thus play a key role in determining the emphasis of a country's climate change responses, which can occupy any position on a sliding scale from policies that focus entirely on one of the three agendas, to policies that focus on one but may result in minor co-benefits for the others, to policies that target genuine synergies, to policies that aim for win-win outcomes across all three agendas (Fisher 2013).

Incentives for the adoption of LCRD frameworks include the opportunity to access international climate funds and engage the private sector, and the potential for a country to demonstrate international leadership in this area (Murphy 2012; Boyle 2013; Fisher et al. 2014). The LCRD policy landscape is still emerging, however, and there are often gaps and failures in implementation at national level. Countries may lack actionable, well-costed plans, or be short of the particular types of expertise they need. Developing a national LCRD strategy can be perceived by actors as an additional, unnecessary burden, especially when they have already established a range of climate change plans (a climate change strategy, climate change action plan or national action plan for adaptation, for instance – see Murphy 2012). Understanding the politics involved is thus crucial to understanding how strategies develop and to assessing the likely outcomes. Chapter 2 of this book looks at national-level LCRD in LDCs in more detail.

Our third point relates specifically to the situation of LDCs, and how the imbalances in resources, opportunities and negotiation capacities that affect these countries also affect their LCRD opportunities. Changes in international climate policy and finance are reshaping incentives, opportunities and barriers to access, but the implications of this for equity and inclusion remain relatively unexplored. There has been some analysis of international funding processes, power relations and institutional issues, and their effect on developing countries. Sharma (2011) and Ayers and Huq (2008), for example, have highlighted how funding processes are often influenced by unbalanced relationships between multilateral funding entities and LDC governments. The expertise of the multilaterals and the control they exert over resources puts them in a position to shape the ways those resources are used. The

LDCs' lack of readiness and capacity to negotiate may further weaken their position. Pre-existing institutional and governance challenges and the complexity of making the transition to LCRD policies mean that many LDCs struggle to balance the demands of growth, poverty and equity. As a result the adoption of an LCRD agenda may in fact only require further trade-offs to be made in low-income countries.

Finally, despite the potential of a constructive use for political economy analysis, for many years it has been more usual to employ it as part of a critical approach. Typically, donors have taken a problem-driven perspective to understanding factors that affect the implementation of their programmes within a country or sector (DFID 2009; Fritz et al. 2009). Analyses undertaken from this perspective have had very little to say about how countries can use an understanding of their internal political economy to make better decisions. In this book we show how political economy factors – actors, ideas and incentives – and the interactions between them can influence national decisions about the design, development and implementation of LCRD. Unpacking these domestic political economy dynamics enables national governments to understand which factors can constrain and which can promote the effectiveness of policy processes and practice. Deployed at appropriate stages of the policy cycle, these findings can be used to improve both the planning and the implementation of LCRD.

Using a political economy lens

As our discussion in the previous section indicates, political economy analysis can be used to analyse the decision-making processes involved in pursuing a national LCRD agenda, and to understand how ideas, power relations and resources interact to shape outcomes (DFID 2009; Tanner and Allouche 2011). This understanding can in turn be put to work to improve the design of LCRD policy and better to understand and manage barriers to implementation, helping to ensure that climate change responses are more effective and equitable. We use the term 'political economy' here to mean the underlying processes by which actors and networks, along with their discourses, knowledge and incentives, shape LCRD outcomes; this definition is based on the work of Tanner and Allouche (2011).

For many decades studies of political economy pursued a 'rational choice' approach, a 'structural' Marxian approach or a 'poststructuralist' approach based on thinking in the areas of economics and political theory, respectively. Our approach brings together these different schools of thought, enabling us to explore more fully the numerous factors influencing climate change and development policies. Tanner and Allouche (2011) and Heirman (2016) characterise this as a 'policy process' approach to understanding the political economy of decision making. In the remainder of this section we describe the conceptual background to this approach.

A rational choice approach reflects actor-centred 'new political economy' thinking, which characterises policies as choices arrived at by assessing the options for distributing economic and political resources (Ndulu et al. 2008). This assumption of 'public choice' rests on the idea that actors make decisions that are consistent with their best interests (Arrow 1951; Bates 1981; Green and Shapiro 1994). On this basis, theorists such as Brennan (2009) posit that policymakers are unlikely to reduce emissions in the absence of binding agreements, as it may not be in their economic interests to do so. This choice-based approach highlights the importance of incentives and resources as key elements in both political and economic outcomes, and their interaction with politics and actors in generating policy directions. Incentives are the perceived rewards and penalties, material and non-material, that motivate actors to behave in certain ways, either enhancing or reducing the attractiveness of a course of action (Giger [1991] 1999; Ostrom et al. 2002). They shape stakeholders' views and their eventual decisions.

A purely actor-centred, rational choice approach thus assumes that actors make the choices that benefit them most. Such approaches look at how individuals, ideas and networks come together to inform policy decisions. This materialist viewpoint has been criticised by a range of scholars, who feel that the 'world cannot be reduced to objective factors which then the rational actor can decide upon' (Riviere 2014). In fact the work of various social psychologists indicates that actors can hold beliefs and ideas that are not rational, given the available evidence (see, for example, Baron and Hannan 1994). These critics point out the importance of the structures, institutions and governance spaces in influencing policy outcomes. This in turn directs us to Marxian structural approaches.

A structural approach characterises policy outcomes as the result of socially constructed and inherited traits – for example rules and norms, and culture and social position, respectively (Roland 2004). This approach focuses more narrowly on the role of the world political order, class structures and vested interests. Its central assumption is that large-scale social systems to a great extent explain the actions of individuals and social groups (Hay 1999; De Canio 2000). An analytical Marxian approach takes this a step further, explaining the relationship between individual agency and social structure (Lebowitz 1998; Mayer 1989). Although these perspectives include a role for policymakers, they do not depend solely on actors' choices (Elster 1982). A purely structural approach assumes that governments make decisions that preserve existing institutions and norms. However, this does not ring true: in terms of the LCRD policy process, the external environment for climate planning and investment and the institutional contexts of key actors undoubtedly play a role in decision making, but individual actors, and their ideas and narratives, also have some influence.

On this basis, in isolation neither of these two approaches – rational choice or structural –allows us to assess the multiple factors that influence LCRD policy-making processes and outcomes.

There have been other combined ways of understanding policy processes. The first is an integrated actor-structural approach that relates political economy to both agency and structure, and continuously engages with the intersection of both (Giddens 2011). It emphasises that both actors and the structures within which they operate influence policy (Rothstein 1968).

The second is a poststructuralist approach that looks at how discourses and ideas motivate policy decisions. Here 'discourses' are 'constructive phenomena shaping the identities and practices of human subjects' (Foucault 1972). We use the term to mean shared ways of thinking, talking or interpreting social and physical phenomena (Dryzek 2000). In some circumstances, these shared understandings or narratives are used by actors, networks or institutions to justify particular choices and actions (Roe 1991; Leach et al. 2010). Analysing discourse entails understanding how language and text are used to produce or reproduce specific outcomes (Luke 1995); how the 'continuous interplay of discourse, political interests and the agency of multiple actors' plays out (Keeley and Scoones 1999). More broadly, Hajer (1995) uses the related concept of 'storylines', which represent the ways actors understand and describe policy issues, and around which consensus emerges; policy options that do not fit in with a consensus storyline are closed off, thus limiting critical reflection. According to this definition, a storyline is a particular narrative on a particular issue, within a broader discourse (that is, a policy discourse might include several storylines).

Recent work on discourses and narratives suggests that no single actor has the resources to bring about change; however, networks of actors with shared beliefs do have this capacity (Newell 2015; Schmitz 2015). Hajer (1995) argues that 'discourse coalitions' form around certain storylines. Actors within these coalitions may not share the same goals, but do have similar understandings of causes and effects, and the range of possible solutions. Schmitz, meanwhile, argues that 'focusing on narrative based alliances is essential for understanding advances and setbacks in sustainability transformations'; analysis of such alliances can provide insight into how shared interests and purpose can influence LCRD policy or create resistance to implementation (Schmitz 2015). The analysis of discourse coalitions is particularly significant to understanding LCRD, since this is an area of policy and debate in which core ideas are highly contested, but nevertheless play an important role in the framing of an agenda and objectives. Discourse coalition, however, remains different from a policy network in which a coalition is looser but is able to influence decisions through the ways ideas are structured. Policy networks can develop when discourses are concretised by exchanging resources. Borzel (1998) describes these policy networks as a 'set of relatively stable relationships which are of non-hierarchical and interdependent nature linking a variety of actors, who share common interests with regards to policy and who exchange resources to pursue these shared interests, acknowledging that cooperation is the best way to achieve common goals'.

In analysing discourse it is important to assess the role of knowledge. A discourse can be held together by knowledge or can shape how knowledge is used. Discourse also has a role in how different forms of knowledge are valued and how they are used in policymaking; knowledge can be obscured or reframed according to actors' beliefs and interests. A discourse analysis can thus reveal how knowledge is used (Atkinson et al. 2010). For Foucault (2002), 'discourse is about the production of knowledge through language.' It can be used to assert 'power and knowledge' or express 'resistance and critique' (Luke 1995). The use of knowledge has been recognised as being particularly political in the context of climate change (see Keeley and Scoones 2003; Leach 2007; Adger 2007), whereby policies are not only linked to scientific evidence but also to the ideation of certain concepts due to interests, which can increase or decrease the importance of the available evidence. A discourse analysis can help us to understand 'how certain meanings have emerged and been framed, while others have been obscured' (Pettenger 2007).

The analytical framework used in this book draws on several of these approaches to examine the political economy of LCRD policy- and decision-making processes:

- The actor-centred rational choice approach, which emphasises the importance of actors, their networks and the underlying incentives that shape processes.
- Structural approaches, which underline the role of the policy landscape, institutions, norms and structures within which decisions are made.
- Discourse-based approaches, which highlight the role of knowledge and discourse coalitions in shaping policies.

Using these perspectives, we can analyse the interaction of actors, coalitions, networks, discourses, incentives and institutions in order to understand more fully the complex political and ideological processes underpinning LCRD policies.

Analytical framework and methods

Following on from the four main gaps we identified in the literature on LCRD policy (see the section on 'Why does the political economy of LCRD matter?'), in this book we focus on four related, but underexplored, questions:

- What role does the domestic political economy of a country play in shaping the planning and implementation of LCRD policies?
- How are countries developing and implementing their LCRD strategies?
- What are the implications for the LDCs?
- How can countries use political economy analysis constructively to improve LCRD planning and outcomes?

Politics of LCRD in the LDCs 13

These questions are addressed in Chapters 3–7, making use of empirical evidence on LCRD policymaking in LDCs, as well as the political economy approach discussed in the previous section. Each of these chapters focuses in detail on one or more aspects of political economy, as follows:

- Chapter 3: the role of discourse in LCRD at national level, and its implications for achieving synergies between climate change mitigation and adaptation, and development;
- Chapter 4: the political economy of CIFs decision making at national level, and how discourse coalitions and incentives influence the outcomes of the programme;
- Chapter 5: the actors, policy networks and institutional factors that have shaped national climate finance systems;
- Chapter 6: the role of incentives throughout the finance chain in delivering climate finance to low-income communities;
- Chapter 7: using political economy analysis to improve decision making within national climate finance systems.

The analytical framework used throughout the book has three dimensions: the political economy; the policy cycle; and the concept of multiple scales (see Figure 1.1).

The three main components of political economy, as shown in the diagram, are actors and networks, discourses and knowledge, and resources and incentives.

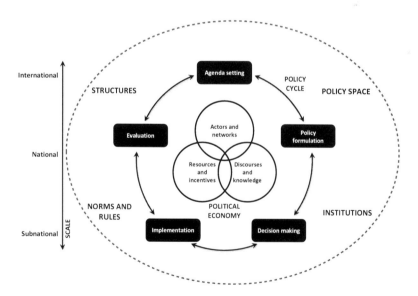

Figure 1.1 Analytical framework used in this book

- **Actors and networks** Actors are the individuals, groups and organisations involved in LCRD policymaking and implementation. They may be international, national or subnational; state or non-state; public or private. They support, compete or cooperate in decisions and actions; their influence varies considerably depending on the nature of their involvement. Our approach characterises actors both as agents of change (an actor-centred approach) and as part of change processes (a structural approach).

  Actors with shared views and interests may also form policy networks around specific issues, coordinating the use of resources in order to achieve their goals (Borzel 1998). We include analyses of how such networks form, and the implications for integrated policymaking and LCRD outcomes.
- **Discourses and storylines** Some chapters follow the approach of Dryzek (2000) and Hajer (1993) and identify discourses and narratives. Others use a more detailed Hajerian storyline analysis, which has four elements: the overarching narrative (or policy discourse); the storylines within that narrative; the coalitions that support particular storylines or discourses; and the role of the narrative in forming policy.
- **Resources and incentives** Incentives are 'the rewards and punishments that are perceived by individuals to be related to their actions and those of others' (Ostrom et al. 2002). Incentives may be material or non-material, and can be categorised as:

  a   policy incentives: legislation; regulations or institutional mandates that support particular discourses and decisions;
  b   economic incentives: resources; funds; technologies; and so on;
  c   knowledge and capacity incentives: the evidence, information and available technical skills that support one decision rather than another;
  d   reputational incentives: the idea that a specific action will enhance the reputation and level of goodwill for the actors or institutions involved;
  e   socioeconomic incentives: positive socioeconomic changes such as improved livelihoods, education or health, or reduced inequality.

Our analysis is aimed at understanding how these components of the political economy – actors, discourses and incentives – interact with existing structures (including the geographic and historical characteristics of the policy space), cultural and social positions, and institutions which are 'the rules of the game in society or, more formally, are the humanly devised constraints that shape human interaction' (North 1990). These formal and informal norms, rules and systems also govern interactions between discourses and behaviour.

Power is an important dimension of a political economy analysis. In this book, however, we have analysed power as embedded within language, discourses and incentives. We particularly look at power in terms of how knowledge is framed and used, and how incentives and resources are garnered to achieve outcomes.

Returning to Figure 1.1, we analyse the three components of political economy at all stages of the policy cycle: agenda setting: policy formulation: decision making: implementation: and evaluation and learning. As the double arrows suggest, however, we do not view these stages as sequential or unidirectional; policymaking is not a straightforward linear process (Howlett and Ramesh 2003).

Finally, our analyses include the dimension of levels or scale. As noted by Tanner and Allouche (2011), scale is a key part of the climate change debate, as it describes the relationship between international negotiations and finance, national planning systems, and local spaces of implementation. We use this concept to assess where ideas, incentives or resources come from and how this impacts on policy processes and outcomes.

Data collection and analysis

The empirical basis for this book involved detailed research in four countries: Bangladesh, Ethiopia, Rwanda and Nepal. These countries were chosen because of their status among the LDCs as early adopters of climate change policies. This meant that they offered opportunities to study a range of approaches to climate planning and relatively developed systems and policies. Local research teams were supported by a small international team responsible for methodology and analysis. Evidence was gathered in the form of policy and institutional analyses, stakeholder interviews and case studies. We also held workshops and action learning groups with relevant stakeholders, in order to present findings and discuss the initial analysis. In each country, research focused on one or two specific issues at national level, and at local level on two case studies concerning how these issues were playing out in practice.

Policy analysis involved examining relevant national policy documents on LCRD planning and finances in order to explore the development of official narratives over time. Stakeholder mapping was used to identify the actors and networks involved, and their role in policy processes. Institutional analysis was also undertaken in order better to understand institutional structures and their implications for policy outcomes.

Semi-structured interviews were carried out with key national stakeholders from four groups: government ministries; research and information services; development partners and international organisations; and the private sector and civil society. The results were analysed in terms of the discourses, narratives and positions of the different actors involved.

Finally, the two case studies carried out in each country focused on the roles of financial intermediaries, as well as exploring how LCRD was being integrated into national planning and finance systems. In some instances where there were no suitable examples of this, we selected cases that dealt with related issues of energy access, renewable energy and development.

Structure of the book and summary of key findings

Using the analytical framework discussed, each empirical chapter examines the political economy dimensions of the LCRD agenda at national and local level. Chapter 2 builds on the framing of LCRD in this chapter to discuss the emergence of LCRD planning in the LDCs, and in particular in our four case study countries. It looks at key climate risks, political factors, institutional mechanisms and finance.

The four chapters that follow form the empirical heart of the book. Chapter 3 looks at how and why Bangladesh, Ethiopia and Rwanda are moving towards LCRD and how discourses and framings around this agenda have emerged and been institutionalised in policy frameworks. Chapter 4 is concerned with how international financial mechanisms interact with national systems, and how political economy dynamics shape LCRD policy and practice. It examines the extent to which two global climate funds have catalysed climate resilient development in Bangladesh, Ethiopia and Nepal. Chapter 5 examines the development of financial instruments, policies and planning systems for LCRD in Ethiopia and Rwanda, and the formation of discourse coalitions to support different options. Chapter 6 draws on examples from Bangladesh and Nepal to show how financial incentives can operate to improve energy access in low-income and rural communities.

In the final two chapters we conclude our arguments. Chapter 7 argues that political economy analysis can be used as a tool at national level to support more effective planning and delivery of LCRD. It recommends ways forward based on the research presented in this book. In Chapter 8 we summarise the book's findings and outline how political economy analysis can inform LCRD policy processes, sharing three core messages:

- **Engaging suitable actors and nurturing policy networks** at different stages of the LCRD policy cycle can provide momentum to an agenda whereby actors are influential, ensure more equitable representation by drawing otherwise marginalised perspectives into the policy process, and improve policy coordination, for example by promoting ownership and knowledge sharing. Chapters 3 and 4 look at the role of actors in setting an agenda and priorities for action, and both suggest that there is a need to include appropriate actors to leverage an agenda, address issues of inclusion and equity, to ensure that diverse views are integrated early on and that actors have a sense of ownership. Chapter 4, in its analysis of two of the CIFs, also shows how engaging actors with a convening authority can support an agenda and take it forward. Chapter 5, meanwhile, illustrates the potential of policy networks for addressing some of

the core challenges of a nascent LCRD agenda, by supporting learning among actors and policy communities.
- **Identifying dominant discourses and emerging coalitions** at an early point can provide insights into the development of support and dissent. Chapters 3 and 4 illustrate the role of discourse, and the particular storylines and coalitions, in shaping the climate change agenda. Chapter 4 also suggests that an early understanding of policy discourse can highlight possible barriers to implementation, in the form of dissenting or conflicting views, as well as helping to ensure that alternative views are not excluded. Where an LCRD agenda is still emerging actors may lack direction and storylines may be supported only by loose discourse coalitions. Supporting these coalitions to develop into more concrete, stable policy networks is one way of moving the agenda forward.
- **Aligning incentives** can help to address the particular challenges of implementation and ensure an equitable and inclusive approach to LCRD. The analysis of renewable energy access interventions in Chapter 6 shows how incentives can be structured to reach the very poorest in society. This involves using high-level policy incentives, recruiting actors with the right capacities and incentives to reach target groups, and tailoring products and services to the needs of potential end users. Chapters 3 and 6 also show how incentives can be aligned to support a particular course of action. Chapter 3, for example, shows that countries' engagement with the LCRD agenda is supported by incentives including the availability of finance, the opportunity to gain the status of a policy leader on the international stage, and the opportunity to support existing national planning priorities.

Taken together, our findings indicate that an understanding of the domestic political economy of LCRD policy is crucial, particularly in the LDCs, which are often excluded from mainstream policy processes. We suggest, therefore, that using political economy analysis to inform policy processes can help to ensure the effectiveness of LCRD efforts and ultimately to improve their outcomes.

Note

1 Ahead of COP21 in Paris in 2015, parties to the UNFCCC agreed to articulate their commitments to and plans for reducing carbon emissions through INDCs. These documents are regarded as countries' individual climate change action plans.

References

Adger, W. N. (2003) Social capital, collective action, and adaptation to climate change. *Economic Geography*, 79(4): 387–404.

Adger, W. N., Agarwal, S., Mirza, M. M. Q., Conde, C., O'Brien, K., Pulhin, J., Pulwarty, R., Smit, B. and Takahashi, K. (2007) Assessment of adaptation practices, options, constraints and capacity. In M. L. Parry, O. F. Canziani, J. P. Palutikof, P. J. van der Linden and C. E. Hanson (eds) *Climate change 2007: impacts, adaptation and vulnerability. Contribution of Working Group II to the fourth assessment report of the Intergovernmental Panel on Climate Change*. Cambridge: Cambridge University Press.

Aldy, J., Ashton, J., Baron, R., Bodansky, D., Charonvitz, S., Dirringer, E., Heller, T., Pershing, J., Shukla, P. R., Tubiana, L., Tudela, F. and Wang, X. (2003) *Beyond Kyoto: advancing the international effort against climate change*. Arlington, VA: Pew Centre on Global Climate Change. Available at http://www.c2es.org/publications/beyond-kyoto-advancing-international-effort-against-climate-change (accessed 26 March 2016).

Arrow, J. (1951) *Social choice and individual values*. New York: Wiley.

Atkinson, R., Held, G. and Jeffares, S. (2010) Theories of discourse and narrative: what do they mean for governance and policy? In R. Atkinson, T. Georgios and K. Zimmermann (eds) *Sustainability in European environmental policy*. Abingdon: Taylor & Francis.

Ayers, J. and Dodman, D. (2010) Climate change adaptation and development I: the state of the debate. *Progress in Development Studies*, 10: 161–168.

Ayers, J. M. and Huq, S. (2008) *Supporting adaptation to climate change: what role for official development assistance?* Conference paper presented at Development's invisible hands: development futures in a changing climate, DSA Annual Conference, London, 8 November. Ayers, J., Anderson, S. and Kaur, N. (2011) Negotiating climate resilience in Nepal. *IDS Bulletin Special Issue: Political Economy of Climate Change*, 42: 70–79.

Bailey, I., MacGill, L., Passey, R. and Compston, H. (2012) The fall (and rise) of carbon pricing in Australia: a political strategy analysis of the carbon pollution reduction scheme. *Environmental Politics*, 21(5): 691–711.

Baron, J. N. and Hannan, M. T. (1994) The impact of economics on contemporary sociology. *Journal of Economic Literature*, 32(3): 1111–1146.

Barrett, S. (2014) Subnational climate justice? Adaptation finance: distribution and climate vulnerability. *World Development*, 58: 130–142.

Bates, R. H. (1981) *Markets and states in tropical Africa: the political basis of agricultural policies*. Berkeley, CA: University of California Press.

Borzel, T. A. (1998) Organizing Babylon: on the different conceptions of policy networks. *Public Administration*, 76: 253–273.

Boyle, J. (2013) *Exploring trends in low-carbon, climate-resilient development*. IISD Paper. Manitoba: IISD. Available at http://www.iisd.org/library/exploring-trends-low-carbon-climate-resilient-development (accessed 29 March 2016).

Boyd, E. and Tompkins, E. L. (2010) *Climate Change: A Beginners Guide*. London: Oneworld.

Bowen A. and Fankauser S. (2011) Low-carbon development for the least developed countries. *World Economics*, 12(1): 145–162.

Bowen, A., Cochrane, S. and Fankhauser, S. (2012) Climate change, adaptation and growth. *Climatic Change* 113(2): 95–106.

Brennan, G. (2009) Climate change: a rational choice politics view. *Australian Journal of Agricultural and Resource Economics*, 53.

Ciplet, D., Roberts, J. T. and Khan, M. (2013) The politics of international climate adaptation funding: divisions in the greenhouse. *Global Environmental Politics*, 13: 49–68.

Curry, J. A. and Webster, P. J. (2011) Climate science and the uncertainty monster. *American Meterological Society*, 175. Available at http://journals.ametsoc.org/doi/pdf/10.1175/2011BAMS3139.1 (accessed 29 March 2016).

De Canio, S. (2000) Beyond Marxist state theory: state autonomy in democratic societies. *Critical Review*, 14: 215–236.

Dessler, A. and Parson, E. A. (2010). *The Science and Politics of Global Climate Change: A Guide to the Debate*. Cambridge: Cambridge University Press.

DFID (2009) *Political economy analysis: how to note*. DFID Practice Paper. Available at http://www.odi.org/sites/odi.org.uk/files/odi-assets/events-documents/3797.pdf (accessed 20 January 2016).

Dodman, D. and Mitlin, D. (2015) The national and local politics of climate change adaptation in Zimbabwe. *Climate and Development*, 7: 223–224.

Dryzek, J. S. (2000) *Deliberative democracy and beyond: liberals, critics, contestations*. Oxford: Oxford University Press.

Dyck, M. G., Soon, W., Baydack, R. K., Legates, D. R., Baliunas, S., Ball, T. F. and Hancock, L. O. (2007) Polar bears of western Hudson Bay and climate change: are warming spring air temperatures the 'survival' control factor? *Ecological Complexity*, 4: 73–84.

Elster, J. (1982) Marxism, functionalism and game theory: the case for methodological individualism. *Theory and Society*, 11: 453–482.

Fisher, S. (2013) *Low-carbon resilient development in the least developed countries*. IIED Issue Paper. Available at http://pubs.iied.org/pdfs/10049IIED.pdf (accessed 29 March 2016).

Fisher, S., Fikreyesus, D., Islam, N., Kalore, M., Kaur, N., Shamshuddoha, M. D., Nash, E., Rai, N., Tesfaye, L. and Rwirahira, J. (2014) *Bringing together the low-carbon and resilience agendas: Bangladesh, Ethiopia, Rwanda*. IIED Working Paper. Available at http://pubs.iied.org/10099IIED (accessed 19 March 2016).

Foucault, M. (1972) *The archeology of knowledge*. London: Tavistock Publications.

Foucault, M. (2002) *The order of things: an archaeology of the human science*. New York: Routledge.

Fritz, V., Kaiser, K. and Levy, B. (2009) *Problem-driven governance and political economy analysis: Good practice framework*. Washington, DC: World Bank.

Giddens, A. (2011) *The politics of climate change*, 2nd edn. Cambridge: Polity Press.

Giger, M. ([1991] 1999) Avoiding the shortcut: moving beyond the use of direct incentives. A review of experiences with the use of incentives in projects for sustainable soil management. *Development and Environment Reports* 17, Centre for Development and Environment, Berne.

Green, D. and Shapiro, I. (1994) *Pathologies of rational choice theory: a critique of applications in political science*. New Haven, CT: Yale University Press.

Goklany, I. M. (2005) A climate policy for the short and medium term: stabilization or adaptation? *Energy and Environment*, 16: 667–680.

Goklany, I. M. (2009) Is climate change the 'defining challenge of our age'? *Energy and Environment*, 20: 279–302.

Hajer, M. A. (1993) Discourse coalitions and the institutionalisation of practice: the case of acid rain in Britain. In F. Fisher and J. Forrester (eds) *The Argumentative Turn in Policy Analysis and Planning*. London: Duke University Press, pp. 43–76.

Hajer, M. A. (1995) *The politics of environmental discourse: ecological modernization and the policy process*. Oxford: Oxford University Press.

Hay, C. (1999) Marxism and the state. In A. Gamble, R. Marsh and T. Tant (eds) *Marxism and social science*. Basingstoke: Macmillan.

Heirman, J. (2016) *The impact of international actors on domestic agricultural policy: a comparison of cocoa and rice in Ghana*. PhD Dissertation. University of Oxford.

Hochstetler, K. and Viola, E. (2012) Brazil and the politics of climate change: beyond the global commons. *Environmental Politics*, 21: 753–771.

Howlett, M. and Ramesh, M. (2003) *Studying Public Policy: Policy Cycles and Policy Subsystems.* Oxford: Oxford University Press.

Illman, J., Halonen, M., Rinne, P., Huq, S. and Tveitdal, S. (2012) *Scoping study on financing adaptation-mitigation synergy activities.* Copenhagen: Nordic Council of Ministers.

IPCC (2007) Summary for policymakers. In M. L. Parry, O. F. Canziani, J. P. Palutikof, P. J. van der Linden and C. E. Hanson (eds) *Climate change 2007: impacts, adaptation and vulnerability. Contribution of Working Group II to the fourth assessment report of the Intergovernmental Panel on Climate Change.* Cambridge: Cambridge University Press, pp. 7–22.

IPCC (2014) Summary for policymakers. In C. B. Field, V. R. Barros, D. J. Dokken, K. J. Mach, M. D. Mastrandrea, T. E. Bilir, M. Chatterjee, K. L. Ebi, Y. O. Estrada, R. C. Genova, B. Girma, E. S. Kissel, A. N. Levy, S. MacCracken, P. R. Mastrandrea, and L. L. White, (eds) *Climate change 2014: impacts, adaptation and vulnerability. Part A: Global and sectoral aspects. Contribution of Working Group II to the fifth assessment report of the Intergovernmental Panel on Climate Change.* Cambridge: Cambridge University Press, pp. 1–32.

Keeley, J. and Scoones, I. (1999) *Understanding environmental policy processes: a review.* IDS Working Paper 89. Available at https://www.ids.ac.uk/files/dmfile/wp89.pdf (accessed 23 March 2016).

Keeley, J. and Scoones, I. (2003) *Understanding Environmental Policy Processes: Cases from Africa.* London: Earthscan, pp. 21–39.

Khan, R. M. (2014) *Toward a binding climate change adaptation regime: a proposed framework.* Abingdon: Routledge.

Klein, R., Eriksen, S. E. H., Naess, L. O., Hammil, A., Tanner, T., Robledo, C. and O'Brien, K. L. (2007) Portfolio screening to support the mainstreaming of adaptation into climate change and development assistance. *Climatic Change*, 84: 23–44.

Leach, M., Scoones, I. and Stirling, A. (2010) *Dynamic sustainabilities: technology, environment, social justice.* London: Earthscan.

Leach M., Bloom G., Ely A., Nightingale P., Scoones I., Shah E. and Smith, A. (2007) *Understanding governance: pathways to sustainability.* STEPS Working Paper 2. Brighton: STEPS Centre.

Lebowitz, M. (1988) 'Analytical Marxism' Marxism? *Science and Society*, 52: 191–214.

Luke, A. (1995) Text and discourse in education: an introduction to critical discourse analysis. In M. W. Apple (ed.) *Review of research in education* Vol. 21. Washington, DC: American Educational Research Association, pp. 3–48.

Luterbacher, U. and Sprinz, D. F. (eds) (2001) *International relations and global climate change.* Cambridge, MA: MIT Press.

McCright, A. M. and Dunlap, R. E. (2011) The politicization of climate change and polarization in the American public's views of global warming, 2001–2010. *Sociological Quarterly*, 52: 155–194.

McGray, H., Hammil, A. and Bradley, R. (2007) *Weathering the storm options for framing adaptation and development.* Washington, DC: World Resources Institute.

Mayer, T. (1989) In defence of analytical Marxism. *Science and Society*, 53: 416–441.

Murphy, D. (2012) *Kenya's climate change action plan subcomponent 1: low carbon climate resilient development.* Presentation at COP18 official side event. Durban, South Africa.

Nakhooda, S. and Norman, M. (2014) *Climate finance: is it making a difference? A review of the effectiveness of multilateral climate funds.* London: Overseas Development Institute. Available at http://www.odi.org/sites/odi.org.uk/files/odi-assets/publications-opinion-files/9359.pdf (accessed 19 March 2016).

Ndulu, B. J., O'Connell, S., Bates, R., Collier, P. and Soludo, C. (eds) (2008) *The political economy of economic growth in Africa, 1960–2000.* Cambridge: Cambridge University Press.

Newell, P. (2015) The politics of green transformations in capitalism. In I. Scoones, M. Leach and P. Newell (eds) *The politics of green transformations.* Abingdon: Routledge.

Newell, P., Phillips, J. and Purohit, P. (eds) (2011) *The political economy of clean development in India: CDM and beyond.* Brighton: IDS.

North, D. (1990) *Institutions, institutional change and economic performance.* Cambridge: Cambridge University Press.

Ostrom, E., Gibson, C., Shivakumar, S. and Andersson, K. (2002) *Aid, incentives, and sustainability. An institutional analysis of development cooperation.* Stockholm: Sida.

Paavola, J. and Adger, N. (2006) Fair adaptation to climate change. *Ecological Economics,* 56(4): 594–609.

Parry, M. L., Canziani, O. F., Palutikof, J. P., van der Linden, P. J. and Hanson, C. E. (eds) (2007) *Climate change 2007: impacts, adaptation and vulnerability. Contribution of Working Group II to the fourth assessment report of the Intergovernmental Panel on Climate Change.* IPCC Report. Cambridge: Cambridge University Press.

Paterson, M. and Grubb, M. (1992) The international politics of climate change. *International Affairs,* 68: 293–310.

Pettenger, M. E. (2007) Introduction: power, knowledge and the social construction of climate change. In M. E. Pettenger (ed.) *The social construction of climate change: power, knowledge, norms, discourses.* Aldershot: Ashgate Publishing, pp. 1–20.

Riviere, L. L. (2014) Towards a constructivist international political economy of climate change. *Issues in Political Economy,* 23: 90–101.

Roe, E. (1991) Development narratives, or making the best of blueprint development. *World Development,* 19(4): 287–300.

Roland, G. (2004) Understanding institutional change: fast-moving and slow-moving institutions. *Studies in Comparative International Development,* 38: 109–131.

Rothstein, R. L. (1968) *Alliances and Small Powers.* New York: Columbia University Press.

Salehyan, I. and Hendrix, C. S. (2010) Science and the international politics of climate change. *Whitehead Journal of Diplomacy and International Relations.* Available at https://www.ciaonet.org/attachments/23685/uploads.

Schmitz, H. (2015) Green transformation: is there a fast track? In I. Scoones, M. Leach and P. Newell (eds) *The politics of green transformations.* Abingdon: Routledge.

Sharma, S. K. (2011) The political economy of climate change governance in the Himalayan region of Asia: a case study of Nepal. *Procedia – Social and Behavioral Sciences,* 14: 129–140.

Sprinz, D. F. and Weiss, M. (2001) Domestic politics and global climate policy. In U. Luterbacher and F. Sprinz (eds) *International relations and global climate change.* London: MIT Press.

Tanner, T. and Allouche, J. (2011) Towards a new political economy of climate change and development. *IDS Bulletin,* 42: 1–14.

Tanner, T. and Mitchell, T. (2008) Poverty in a changing climate. *IDS Bulletin,* 39: 4.

2 National and international experiences of low carbon resilient development

Susannah Fisher, Dave Steinbach, Costanza Poggi and Saleemul Huq

Introduction

In this chapter we look in more detail at the LCRD planning paradigm introduced in Chapter 1. We start by exploring how this paradigm has evolved at international level and how it has been financed in the LDCs. We go on to consider the particular questions and challenges it throws up for LDCs at national level that have emerged from existing research. We then analyse in more detail the four countries under study in this book: Bangladesh, Ethiopia, Rwanda and Nepal. In each case, we briefly outline the internal climate change planning context – its history and institutional and funding arrangements – before going on to identify the key features of its LCRD planning. This draws together information from across the authors' research (and specifically from Fisher and Slaney 2013; Fikreyesus et al. 2014; Rai et al. 2014 and Nash and Ngabitsinze 2014).

Our analysis of the international context, national research findings and four national cases leads us to suggest that while international political developments and finance have given national governments the initial push to engage with the LCRD agenda, the dynamics of these countries' internal political economy are also playing a significant role in shaping their plans and actions. There are a range of challenges with national planning in this area and governments are putting in place policy frameworks, financial and institutional mechanisms to address particular issues identified. This provides the context for the later chapters of this book which explore the political economy of these in more detail.

We begin, though, by outlining how the LCRD agenda has emerged through international negotiations and climate funds.

The evolution of the LCRD paradigm: mitigation, adaptation and finance

Since its establishment in 1994, the UNFCCC has been the primary forum for negotiating a climate change agreement at international level. Each year a COP is convened, at which representatives from the governments of all countries that are signatories to the Convention meet to negotiate an international response. Initially, the focus of UNFCCC negotiations was on stabilising greenhouse gas emissions to prevent dangerous anthropogenic interference with the climate system, reflecting the recognition that greater mitigation would reduce the need for adaptation in the long term (Schipper 2006).

The Kyoto Protocol, signed in 1997, committed 37 industrialised countries and the 15 original members of the European Community to legally binding emissions reduction targets for the main greenhouse gases. Mitigation continued to be the main pillar of international climate policy following this agreement. A number of intersecting mitigation themes and agendas emerged, along with a variety of sector-specific initiatives.

Over time, however, the focus on climate change mitigation came to be challenged by those advocating for adaptation measures. Factors behind this shift included the perceived lack of success at COP6, which took place in 2000 in The Hague, Netherlands, in establishing Kyoto Protocol targets and implementation, followed by the complete withdrawal of the United States from the Protocol, illustrating the political and practical challenges of mitigation (Schipper 2006). In addition, the IPCC's *Third Assessment Report*, released in 2001, made it clear that mitigation efforts were not going to prevent climate change impacts, and that these impacts would particularly affect the poorest and least developed countries (Ayers and Dodman 2010).

This new focus on climate change impacts and vulnerability led to policy responses aimed at building the resilience of social, economic and environmental systems and improving their capacity to adapt to climate change. The Least Developed Countries Work Programme was established by the Marrakesh Accords, agreed at COP7 in 2001. This launched several adaptation initiatives and bodies. In particular, it initiated the preparation by LDCs of National Adaptation Programmes of Action (NAPAs). These are plans identifying priorities for immediate and urgent adaptation actions that are eligible for funding from the Least Developed Countries Fund (LDCF), managed through the Global Environment Facility.

Under the terms of the 2010 Cancun Adaptation Framework, adaptation was formally recognised as an equal pillar, alongside mitigation, of a future global climate agreement (UNFCCC 2011). The Framework also established a National Adaptation Plan (NAP) process, which supported countries to address their medium- and long-term adaptation needs. Developing countries have also been invited to submit NAMAs, outlining national plans to reduce emissions relative to business-as-usual levels in 2020, which will be supported by technology, capacity and finance. In preparation for COP21 in 2015 each of the states parties were asked to submit an INDC, a single document reporting their national mitigation and adaptation commitments. These documents were often used to communicate existing plans developed through the NAPA, NAMA and other national planning processes. (See Table 2.1 for a summary of the main UNFCCC plans affecting LDCs.)

Table 2.1 Main UNFCCC planning frameworks affecting LDCs

Plan	Main aim	Status in LDCs
National Adaptation Programme of Action (NAPA)	Urgent and immediate adaptation needs	Almost all have developed NAPAs; however, many not fully funded in implementation
Nationally Appropriate Mitigation Action (NAMA)	Actions to reduce emissions that can receive financial and other support	Some have submitted plans, of which a subset have been funded
National Adaptation Plan (NAP)	Medium- and long-term adaptation needs	LDCs in the process of developing these at the time of writing

As international policy responses to climate change have developed, there have been increasing efforts to provide additional funding for their implementation in developing countries. Finance has been made available through multilateral funds (managed both within and outside the UNFCCC), bilateral sources (development assistance), domestic sources and the private sector, and this has led to a reconfiguring of relationships at international and national level, as discussed in Chapter 1. Here we give more detail about key funds and concerns regarding climate finance for the LDCs.

Multilateral funds are the newest of these funding channels, and UNFCCC funds are important sources of finance for national climate change planning in the LDCs. Two examples are the Special Climate Change Fund (SCCF) and the LDCF, which focus on adaptation in developing countries and in LDCs, respectively. They are managed by the Global Environmental Facility, a trustee of the World Bank, while projects are implemented by multilateral institutions such as the UN Development Programme (UNDP) or the Asian or African Development Banks. Since 2009 the Adaptation Fund, established via the UNFCCC and the Kyoto Protocol, has also funded adaptation activities in developing countries. It is financed by proceeds from Clean Development Mechanism projects and donor pledges.

The GCF was formally established at COP16 in Cancún, Mexico. It is expected to be the primary international channel for financing climate change responses in developing countries, and forms part of developed countries' commitment to providing US $100 billion per year for this purpose by 2020. The Fund selected its first eight projects in late 2015 (Green Climate Fund 2015).

For LDCs, there are several key issues relating to financing LCRD: the scale and distribution of finance; the accessibility of finance; and how to pursue both the mitigation and adaptation agendas.

With regard to the scale and distribution of finance, Nakhooda and Norman show that while funding for adaptation has targeted LDCs to some extent, 'the volume of finance that LDCs have received is modest in absolute terms, reflecting the small size of these funds. And we should note that not all LDCs have received adaptation finance' (2014: 39). The UNFCCC's LDCF is the only fund specific to LDCs, and until funds were pledged at the 2015 Paris COP it was empty, leaving 35 projects stalled in the pipeline (Uprety 2015; LDC Group Chair 2015). The LDCF operates on a system of voluntary contributions, and this, along with recent experience of low levels of finance for the fund, raises fundamental questions about the scale and sustainability of finance that LDCs can access (Tenzing et al. 2015). The GCF has received significant pledges and commitments from donor countries, but it remains to be seen whether these levels of financing can be sustained.

Modes of access to funds vary and have been a source of controversy. Two UNFCCC funds – the GCF and the Adaptation Fund – allow direct access to accredited government entities. Two others – the SCCF and the LDCF – allow only indirect national access, managed through multilateral, international and bilateral entities (Bird et al. 2011). This indirect access has been controversial owing to the high costs and transaction times needed to work with these third parties (Uprety 2015). Proponents for the LDC Group have asked for direct access to the LDCF, to simplify and speed up the process, although it also recognised that accreditation for direct access to the other funds has been challenging for LDCs so far (Uprety 2015; Tenzing et al. 2015).

Funding for mitigation and adaptation remains unbalanced, with most of it going to mitigation efforts in emerging economies. The 2009 Copenhagen Accord made some attempt to correct this, but recent trends suggest that little progress has been made (Nakhooda and Norman 2014). Theoretically, LCRD policies can be used to channel finance through national planning processes for joint adaptation and mitigation objectives. However, there are few examples of programmes aimed at achieving such synergies. In the absence of clear, costed plans, and given that international and domestic financing systems are not geared towards this kind of integrated approach, this means that in most countries the task of finding synergies within LCRD is at an early stage. The investment priorities for the GCF provide opportunities for financing LCRD programmes – under headings such as transforming energy generation and access, and encouraging low-emission and climate resilient agriculture – but it remains to be seen how these will be operationalised.

In summary, international policies have prompted national governments to address immediate adaptation needs in NAPAs, to consider mitigation efforts in NAMAs, to reflect on longer-term needs in NAPs, as well as to report on their mitigation and adaptation commitments in INDCs. The opportunity to access international climate finance from both bilateral and multilateral sources creates a variety of incentives that may inform countries' climate change planning, and reconfigures national and international relationships. In the next section we consider how national LCRD policies have emerged, and the challenges associated with them.

Planning for LCRD at national level

As well as developing NAPAs, NAMAs and NAPs – all of which are internationally driven – some LDC governments have created separate national LCRD plans. Although these plans build on earlier policy documents they also reflect national development priorities, and many also aim to bring climate change policy into the main national planning arena. Furthermore, some address issues of adaptation and mitigation jointly (Fisher 2013). A 2015 survey of LDC climate planning found that 20 out of 48 LDCs had drafted a climate change policy or strategy document, many of which contained at least some mention of both adaptation and mitigation agendas (Fisher and Mohun 2015).

These policies mark an internal shift in these countries, away from treating climate change as an issue to be dealt with by environment ministries, and limited to a matter for UNFCCC negotiations, to placing it in the hands of powerful sectoral line ministries such as agriculture and finance where it becomes a core policy concern. Some governments are also trying to finance their plans through the national budget, recognising that international finance will be inadequate.

LCRD planning and implementation is a challenging public policy issue. As outlined in Chapter 1, it involves multiple actors from different policy communities (i.e. it is cross-sectoral), overlapping responsibilities and jurisdictions, and a contested knowledge and inputs from different communities of 'experts'. The adaptation, mitigation and development agendas overlap and areas of intersection may give rise either to synergies or to trade-offs; this carries the risk of making maladaptive short-term decisions that impact negatively on long-term outcomes.

Research in this area is in the early stages but there are some notable exceptions which we explore here using illustrative national examples to draw out lessons for this book. We then go on to look at how institutions and frameworks have emerged in our four country case studies to deal with these challenges.

Relationships within and between scales

LCRD operates at multiple scales, in terms of areas of responsibility, finance, support and implementation.

A disconnect has often been identified between international climate initiatives and national implementation. The Pilot Program for Climate

Resilience (PPCR, one of the CIFs) has allowed researchers to explore issues of climate planning in the absence of widespread national policy-making in this area, particularly the relationships between national and international actors. The programme provides loans, grants and technical assistance to mainstream adaptation into development planning, and the implementation of PPCR projects at local level has given rise to a variety of ideological, political and social tensions. Seballos and Kreft (2011) identify the involvement of the World Bank and other multilateral development banks in implementing the PPCR as an early point of contention. They also highlight criticisms of the potential for PPCR loans to increase the debt burden on vulnerable countries, the possibility of undermining the activities of the UNFCCC's Adaptation Fund, a lack of harmonisation with country-level initiatives, and poor engagement with civil society.

The way in which national and international relationships develop depends upon which key figures are involved at different stages, and upon their relative positions and power, and this has implications for the climate-development agenda and national ownership. In their analysis of cross-country evidence from developing countries Tanner and Allouche (2011) identify three findings concerning power and inclusion. First, a sense of urgency and time constraints around international programmes affect the development of what are complex initiatives, creating barriers to meaningful and wide consultation. Second, existing channels of accountability and inclusion are typically either issue- or sector-driven, which can mean that climate change considerations, cutting across these issues and sectors, are excluded. Third, the framing of international actors and national governments as the primary agents for climate change action tends to limit the involvement of other actors. A recurring tension has been identified in other research around perceived top-down donor-driven projects, which have allowed for limited input of relevant stakeholders in the design and implementation processes (Shankland and Chambote 2011; Alam et al. 2011; Ayers et al. 2011). This has raised questions about the ownership of local climate planning (Alam et al. 2011). However, even without these issues researchers have also identified a lack of national leadership and significant contestation within LCRD policy processes, with decisions being explained by path dependencies and institutional inertia rather than by leadership and ideas (Tanner and Allouche 2011).

Unofficial policy spaces also need to be considered. These are occupied by civil society and non-state and subnational actors, and bottom-up movements and governance mechanisms that impact on national and international processes (Bulkeley and Betsill 2005; Pattberg and Stripple 2008; Fisher 2014). In order to understand the policy process more fully, we need to ensure that the scope of our enquiry is wide enough to include actors like these, and to consider how actions and decisions in one policy space can influence another.

Trade-offs and win-wins

Within an LCRD agenda overlaps between the adaptation, mitigation and development agendas can have either positive or negative consequences. It may be that one agenda impacts negatively on another, meaning that the costs and benefits need to be managed (trade-offs); equally there may be mutual synergies between agendas (co-benefits) and policies that together address adaptation, mitigation and development (known as a triple win). There has been very little empirical examination of these areas of intersection (Suckall et al. 2014). There is also a complex and contested relationship between growth, poverty and equity in the LCRD agenda. Ellis et al. describe the complex relationship thus: 'there will be winners and losers, and trade-offs between social, economic and environmental goals and between long-term and short-term benefits' (2013: 11). Renewable and low carbon energy sources are a fundamental part of LCRD, particularly in developing countries, where the need to tackle energy poverty and promote energy security coincides with mitigation objectives. As such, it is necessary to ask how and why to link the mitigation, adaptation and development agendas within a national context, why specific technologies and renewable energy sources are chosen over others, and the implications that this has for different communities and national development.

The following examples look at how this intersection has played out in different contexts. The energy sector in Kenya provides an interesting example of the potential and limits of LCRD as national policy priorities for cheap secure electricity tend to become aligned with some renewable energy technologies (although the recent discovery of oil may unsettle this). Newell et al. have analysed trade-offs between renewable energy development and pro-poor energy initiatives in Kenya and argue that while incentives align behind supporting geothermal energy to improve access to electricity, 'access to electricity is only one dimension of energy poverty. Cheap electricity does not necessarily mean pro-poor electricity or that all energy service needs are served. The drivers of policy have not, on the whole, been concerns for pro-poor energy access, but rather concerns around energy security and the competitiveness of industry in Kenya' (2014: 32).

Local examples also show that the intersections need to be critically considered as they may have negative consequences. Tompkins et al. use an analysis of coastal management in Belize, Ghana, Kenya and Vietnam to show that, while some policies leverage triple wins, others are 'creating development losses, maladaptation and worsening emissions' (2013: 16). The authors argue that the simplified depiction of win-wins can hide trade-offs and regrets. 'Without a strong evidence base there is a risk that the development community could invest in policies that create triple wins with regrets at the expense of more effective policies that might only deliver co-benefits but with no-regrets' (ibid.: 17). Suckall et al. (2014) analyse four case studies in sub-Saharan Africa for triple wins and argue that while triple wins are possible, there are often no explicit criteria to evaluate each of the three areas and a lack of data on whether or not they have been achieved.

What any of these synergies might be is also a matter of contextual factors. Tanner et al. (2014) provide two examples that illustrate the challenges of achieving triple wins in practice, as well as the need for externally driven, low carbon development strategies to consider local dynamics. These examples, both from Ghana, relate to initiatives to remove subsidies from premix fuel for artisanal fishing and to protect mangrove forests. Both were problematic, because although they offered considerable potential for reducing emissions and conservation, they would also have a negative impact on the livelihoods of people relying on fishing and wood trading. There was little trust that appropriate compensation or alternative and more sustainable livelihoods would be provided. By examining the multitude of actors, interests and agendas (both national and international) driving the low carbon agenda in Ghana, as well as the current state of the fishing sector, the authors demonstrate that context and competition are key factors in who wins and who loses from low carbon development strategies, and how the relationship between the three LCRD agendas is more fluid and complex than theory might suggest.

Political drivers

Numerous external and internal influences determine how and why countries choose to implement LCRD (Ellis et al. 2013). They range from reducing the cost of vulnerability to climate change and reducing poverty, to 'leapfrogging' to using new technologies and tapping into climate finance. The level of influence of particular incentives varies in different countries, depending on national priorities and goals, and levels of political will for linking the adaptation, mitigation and development agendas (Lockwood and Cameron 2012).

In Ghana, for example, mitigation is a key driver behind policymakers' involvement in LCRD planning – more so than poverty reduction and economic growth – whereas in Ethiopia it is the goal of becoming a middle-income country by 2025 (Lockwood and Cameron 2012). In countries such as Bangladesh and Nepal high levels of vulnerability to extreme weather and natural disasters are central considerations (Ellis et al. 2013). Other common drivers include the potential of renewable energy to improve energy security and resource efficiency for sustained future growth, and strong government leadership.

Ellis et al. (2013) also identify a set of barriers for an LCRD agenda. These are costs associated with change; interest groups opposed to change; lack of awareness or trusted information about uncertainties, risks, opportunities and trade-offs; short-termism; lack of state capacity to respond to and implement strategies; institutional and technological constraints; and lack of clarity about roles and responsibilities.

Lessons

In summary, we can draw several key lessons. First, current evidence shows that relationships between actors at different scales such as the multilateral banks and national actors have been contested, and questions of ownership in early climate programmes are shown to be important. It is important to analyse these early findings as LCRD planning moves into national processes. Second, we can see how particular interests may lead to government preferences for particular energy sources and the role of context in who wins or loses in low carbon development or triple wins. This highlights that this area of intersection between agendas is an important one on which to develop more evidence, and how this plays out in different contexts will be an important empirical area to explore. Finally, a wealth of national political factors and potential barriers have been identified so far, but what has been less explored is how these factors then influence implementation and how some countries seek to overcome these barriers.

We now go on to look at how four LDCs have started to address LCRD in their national planning better to understand current policy, finance and institutional frameworks that might address these challenges.

Case studies

Our analysis in this book focuses on Bangladesh, Ethiopia, Rwanda and Nepal. Of the nine LDCs that had national LCRD plans at the time we began our research (Fisher 2013), these four represent a variety of experiences, in terms of the climate risks they face, their overall approach to climate change policies, their financing arrangements, and their politics and political systems. This allows us to explore how LCRD agendas have emerged in different contexts, which financial and policy systems have been used to support particular objectives, and the underlying political economy factors involved.

We explore these questions in more detail in Chapters 3–7. First, however, in this chapter we outline the characteristics of LCRD policies in the four case study countries, including the national contexts within which they have emerged.

Bangladesh

Climate risks

Together Bangladesh's low-lying position, exposure to extreme weather events and its high population growth rate make it particularly vulnerable to climate change (Rai et al. 2014). High population density and dependency on agriculture and natural environment-based livelihoods are also factors, with the main risks coming from flooding, tropical cyclones, storm surges and drought. The IPCC has estimated that 17 per cent of Bangladesh's land area will have been claimed by rising seas by 2050, forcing 20 million people to relocate (Pervin et al. 2013). Bangladesh's Ministry of Environment and Forests (MOEF) has calculated that the increased frequency and intensity of 'natural' disasters linked to climate change have already cost around 1.5 per cent of gross domestic product (GDP) (Rai et al. 2014).

Political will and policy responses

Over the last decade, concern about the likely impact of climate change on development in Bangladesh has become increasingly widespread. This concern is now reflected in climate change agendas and discourses at international level (Schipper 2006). In 2005, in line with international requirements, the government prepared its NAPA, which identified the urgent actions needed.

Bangladesh's domestic climate change policies were given impetus by Cyclone Sidr, which struck in 2007 killing over 3,000 people and causing direct economic losses of US $1.7 billion. In the same year, monsoon flooding caused extensive damage. Following Sidr, however, both the government of Bangladesh and international stakeholders felt that there was a need for a more long-term strategy than that provided by the NAPA (early discussions, held in 2008, were referred to as 'post-Sidr planning' – see Alam et al. 2011). The result was the development of the Bangladesh Climate Change Strategic Action Plan (BCCSAP).

From the outset, the BCCSAP was far more ambitious than the initial NAPA: it recognised the need for both mitigation and adaptation, and emphasised the government's willingness to follow a low carbon development pathway. However, civil society actors in Bangladesh have been active in promoting the idea that Bangladesh should focus on adaptation rather than mitigation.

Its aim was to facilitate medium- and long-term low carbon and climate resilient development. It comprised two phases spanning ten years, 2009–13 and 2013–18, and six thematic areas: food security, social protection and health; comprehensive disaster management; infrastructure development; research and knowledge management; mitigation and low carbon development; and capacity building and institutional development. It gave rise to a total of 44 programmes, designed to deliver 145 specific actions between them. In 2009 the NAPA was also expanded to include 45 adaptation measures, with 18 immediate and medium-term priorities. In addition, the National Planning Committee has also begun to integrate climate change into longer development planning cycles (Pervin et al. 2013).

The MoEF is the key ministry responsible for climate change-related matters in Bangladesh. Climate Change Cells have been set up within ministries, along with a Climate Change Unit within the MoEF, to coordinate activities. A Department of Climate Change is also being developed. There is a National Environment Committee headed by the prime minister, for strategic guidance and oversight, and a National Steering Committee on climate change chaired by the Minister of Environment and Forests, which aims to harmonise all climate-related activities in Bangladesh (Rai et al. 2014).

Finance

Two climate change funds have been established in Bangladesh: the Bangladesh Climate Change Trust Fund (BCCTF), a domestic government fund, and the Bangladesh Climate Change Resilience Fund (BCCRF). The latter evolved from a multi-donor trust fund and is led by the national government in collaboration with the World Bank and contributing development partners, who have roles on the governing council and management committee.

The BCCTF uses a competitive bidding process to provide finance for projects lasting one to two years. By September 2013 it had funded 139 projects worth a total of US $200 million, all implemented by government departments or non-governmental organisations (NGOs) (Rai et al. 2014). Of this funding, 44 per cent has been spent on the national action plan's (BCCSAP) 'infrastructure development' theme, with less than two per cent directed towards capacity building and institutional development (Pervin et al. 2013). The BCCTF has a board of trustees and a technical committee with representation from 12 ministries, as well as two sub-technical committees of experts.

The BCCRF operates in a very similar manner to the BCCTF in terms of its governance, project selection and implementation partners. The major differences are that it is managed by the World Bank rather than the government of Bangladesh and international donors sit on its governing council. To meet the criteria for funding from this source, projects should require US $15–25 million, be demand-driven and have three-year timelines with a possible one-year extension. Additionally, proposals coming from existing project units are given particular weight during selection and prioritisation.

Box 2.1 Main features of climate change planning in Bangladesh

Policies

- Bangladesh Climate Change Strategic Action Plan (BCCSAP)
- Range of other relevant plans
- Commitment to mainstreaming

Underlying political and climatic factors

- Cyclones and coastal damage
- More political support for adaptation than for mitigation

- Involvement of ministers on governing bodies of national climate change funds (BCCTF and BCCRF)

Cross-sectoral mechanisms for LCRD

- National Environment Committee and National Steering Committee for climate change
- BCCTF board of trustees and technical committee with 16 members from a variety of sectors
- Governing council and technical committee of the BCCRF

Finance

- BCCTF (Bangladesh Climate Change Trust Fund) – board made up of government stakeholders, including ministers
- BCCRF (Bangladesh Climate Change Resilience Fund) – council includes representatives from government and donors

Implementation

- Implementation through application to the BCCTF and the BCCRF, including by civil society and private sector organisations
- Climate Change Unit within the Ministry of Environment and Forests, the Department of Climate Change, being developed
- Climate Change Cells within other ministries

Ethiopia

Climate risks

Ethiopia has long dealt with the impacts of variable and extreme weather, which has resulted in well-publicised famines and food shortages. Both the population and the economy are highly sensitive to changes in climate: approximately 83 per cent of its people depend directly on agriculture for their livelihoods, with many more reliant on agriculture-related industries (Fikreyesus et al. 2014). Agriculture is predominantly rain-fed and as such is very vulnerable to extreme weather events like droughts (Evans 2012).

Projected climate change impacts, including greater variability in precipitation patterns and higher temperatures, will see the frequency and intensity of these extreme weather events increase. These changes are likely to compound the effects of existing drivers of poverty in Ethiopia. A variety of the country's national planning documents highlight specific livelihoods, agro-ecological systems and sectors that are extremely vulnerable to climate change. According to its NAPA, the sectors likely to be most negatively affected are agriculture, water resources and human health (Fikreyesus et al. 2014).

Political will and policy responses

Ethiopia's erstwhile prime minister, Meles Zenawi, launched the Climate Resilient Green Economy (CRGE) Vision in 2011, demonstrating a high level of political engagement (Jones and Carabine 2013). Droughts and food security have been high-profile national and international issues that may also provide incentive for national action (Bass et al. 2013).

The CRGE vision is to build a climate resilient green economy by 2025, pursuing Ethiopia's development objective of achieving middle-income status by the same date by transforming development planning, investments and outcomes. This vision is encapsulated in three main CRGE objectives: reducing greenhouse gas emissions across sectors; reducing vulnerability to climate change; and ensuring economic growth (Fikreyesus et al. 2014). It is supported by two further national strategies: the Green Economy Strategy, launched alongside the CRGE initiative in 2011; and the Climate Resilient Strategy, which was in draft form at the time of writing.

Responsibility for coordinating Ethiopia's CRGE planning lies with the MOFE. The Ministry of Finance and Economic Development (MOFED) is responsible for its financial aspects, while a specific Climate Resilient Green Economy Inter-Ministerial Steering Committee provides oversight. The latter was set up by the Prime Minister's Office to give high-level policy direction, and is made up of ministers and senior officials from each of the ministries contributing to the CRGE. There is also a technical committee and sub-technical committees for specific issues (Bass et al. 2013), and plans exist for CRGE units within ministries and implementing entities.

Ethiopia's development objectives are set out in national Growth and Transformation Plans (GTPs), released every five years. These play a significant role in shaping the interventions of federal and regional bodies throughout the country. A key element that distinguishes the GTPs from previous policy frameworks for Ethiopia is their focus on climate change responses, which are addressed as a strategic priority (termed 'environment and climate change' in the documentation) that cuts across other areas. This makes building a green economy one of the key policies for developing long-term sustainability, and the Plans include objectives, targets and implementation strategies for achieving this.

Finance

A national CRGE Facility has been set up to receive climate funds and finance aspects of the CRGE initiative, disbursing funds based on its priorities. This fund is now operational: to date it has been capitalised by the United Kingdom (UK) and Austria, and has disbursed some fast-track finance to projects. A results framework has also been put in place to measure the country's progress towards its CRGE objectives. Implementing entities such as line ministries and regional bodies can apply for funds from the Facility, as can non-state actors known as 'executing entities'.

> **Box 2.2 Main features of climate change planning in Ethiopia**
>
> **Policies**
>
> - Climate Resilient Green Economy (CRGE) Vision and Strategies
> - Mainstreaming climate change responses within national development plans
>
> **Underlying political and climatic factors**
>
> - High-profile droughts and food security problems
> - Prime Minister's involvement at launch of CRGE
> - Ministerial committee chaired by Prime Minister's Office and including key government ministers
>
> **Cross-sectoral mechanisms for LCRD**
>
> - Technical committees responsible for preparing Green Economy and Climate Resilience Strategies
> - CRGE ministerial committee and finance team
>
> **Finance**
>
> - CRGE Facility – governed by a ministerial committee and the CRGE management committee
>
> **Implementation**
>
> - Line ministries and regional government ('implementing entities') and non-state actors ('executing entities') through access to CRGE Facility funds
> - CRGE unit in each implementing entity

Rwanda

Climate risks

Rwanda has been experiencing annual temperature increases higher than the global average since 1970 (Government of Rwanda 2011), and although rainfall has declined overall during the past decade, extreme rainfall events have become more frequent and intense. These events and their associated consequences – such as landslides and soil erosion – have resulted in fatalities and injuries, population displacements, building and infrastructure damage, and crop failure.

Climate change predictions are that by the 2050s the increase in average temperature in Rwanda will be in the range of 1.5°C to 3°C. Changes in precipitation are more uncertain, though all climate models indicate that there will be changes, with most pointing to an increase in average annual rainfall (typically of about 10 per cent) with particular increases in September to November; some models also predict less rainfall for some months. Projections for extreme events such as floods and droughts (rather than annual averages) is more variable still; however, many models indicate an intensification of heavy rainfall during the wet season, signalling greater flood risk (SEI 2009).

Any changes in rainfall patterns are likely to have a marked impact on Rwanda's economy: rain-fed agriculture employs approximately 80 per cent of the workforce and contributes 35 per cent of GDP (Government of Rwanda 2013), while hydropower currently supplies half of the country's electricity needs and will remain significant in future, supplying around one-third in the longer term (African Development Bank Group 2013).

Political will and policy responses

The government of Rwanda completed its NAPA in 2006. Subsequently, there has been a significant commitment to mainstreaming climate change responses into development planning in Rwanda, along with a commitment to green growth. This has been reflected in policy, for example in NAPA priority actions and in the consolidation of plans within Sector Strategic Plans. The NAPA also informed the development, in 2010–11, of the country's first integrated LCRD plan, the National Strategy for Climate Change and Low Carbon Development (NSCCLCD).

This strategy was the result of a collaborative effort between the government of Rwanda, the University of Oxford's Smith School of Enterprise and Environment, the UK government's DFID-Rwanda and the Climate and Development Knowledge Network and had high-level support from the country's president, Paul Kagame (Ellis et al. 2013).

Its aims are to guide national policy and planning in an integrated way, to mainstream climate change responses in all sectors of the economy and to enable Rwanda to access related international funding. Building on existing climate change and development strategies, it represents a holistic approach to socioeconomic development (see Nash and Ngabitsinze 2014), aimed at helping Rwanda to leapfrog outdated technologies and ineffective development pathways (Government of Rwanda 2011).

The vision set out in the strategy is of a developed low carbon, climate resilient Rwandan economy by 2050. This is based on achieving three strategic objectives:

- energy security and a low carbon energy supply that supports the development of green industry and services, and avoids deforestation;
- sustainable land use and water resource management that results in food security, appropriate urban development, and preservation of biodiversity and ecosystem services;
- social protection, improved health and disaster risk reduction in order to reduce vulnerability to climate change impacts.

In turn, this strategy informed the development, in 2013, of the Economic and Development Poverty Reduction Strategy II (EDPRS 2, the latest of Rwanda's national strategic plans). This makes further progress in mainstreaming LCRD in development planning, particularly in relation to integrated land use, sustainable small-scale energy installations in rural areas and low carbon urban systems (Nash and Ngabitsinze 2014). Some key sectors, including agriculture, have also begun to address this mainstreaming process in separate sectoral strategies.

Two of the key ministries for LCRD in Rwanda are the Ministry of Natural Resources (MINIRENA) and the Ministry of Finance and Economic Planning (MINECOFIN). The Rwanda Environment Management Authority (REMA) operates under MINIRENA, and facilitates implementation of national environmental policy and legislation. REMA has created the Climate Change and International Obligations Unit (CCIOU) which, among a number of functions, is responsible for coordinating the preparation and implementation of policy, strategy and regulatory frameworks, and instruments relating to climate change mitigation and adaptation (Nash and Ngabitsinze 2014). There are also plans for having nominated individuals within all line ministries to lead Rwanda's LCRD agenda.

Finance

The Rwandan government has pooled domestic and external financial resources in a basket fund known as the Rwanda Environment and Climate Change Fund, or FONERWA (see Government of Rwanda 2012). FONERWA is a demand-based fund, meaning that funding decisions are made in response to proposals from project promoters. It disburses climate finance both to the public sector, including government agencies and districts – 10 per cent of funds are earmarked for districts – and to the private sector, which has access to 20 per cent of the total.

Disbursement decisions are guided by FONERWA's investment priorities, which align with the government's broad priorities for the environment, climate change and development, as well as with the national priorities reflected in sector investment plans, and the objectives of the National Strategy and the EDPRS 2.

> **Box 2.3 Main features of climate change planning in Rwanda**
>
> **Policies**
>
> - National Strategy for Climate Change and Low-Carbon Development (NSCCLCD)
> - Mainstreaming climate change responses within national development plans
>
> **Underlying political and climatic factors**
>
> - Increasingly severe climate impacts
> - President's involvement
>
> **Cross-sectoral mechanisms for LCRD**
>
> - MINIRENA – committee on environment and climate change
>
> **Finance**
>
> - FONERWA – national basket fund for environment and climate change
>
> **Implementation**
>
> - Nominated individuals within line ministries
> - Implementation by government line ministries, districts and the private sector, using funds for mainstreaming environment and climate change responses within national planning

Nepal

Climate risks

Nepal has a complex climate due to huge variation in land elevation – a distance of less than 200 kilometres spans the subtropical lowland regions and the Himalayan high mountains (where temperatures remain well below zero throughout winter) – as well as the major influences of the Himalayas and the South Asian monsoon (Government of Nepal 2010). This complexity means that making climate projections for Nepal is extremely challenging. Although the various models all suggest that the country will experience strong warming trends, they differ significantly in the temperatures they predict. Projections for precipitation patterns are still more uncertain: there is no clear trend for annual precipitation, with some models suggesting an increase, others a decrease. Future changes in rainfall variability and extremes are similarly very unclear.

Nepal's economy and the livelihoods and well-being of its people are heavily dependent on its climate. A large proportion of the country's GDP is associated with climate-sensitive activities, particularly agriculture, which is dominated by smallholders and rain-fed production. It is therefore greatly affected by variations in rainfall, extremes of drought, floods, landslides, heat and cold stress, hail and snowfall. Glacial lake outburst floods are also a particular risk (Government of Nepal 2010).

Political will and policy responses

The process of developing a new constitution has slowed policy development in Nepal during the period of our research (Haviland 2015), but a number of the institutional mechanisms for LCRD have high-level engagement from political figures; these include the country's Climate Change Council and its Multi-stakeholder Committee.

Nepal completed its NAPA in 2010, identifying six priority areas for intervention: agriculture and food security; forests and biodiversity; water resources and energy; climate-induced disasters; public health; and urban settlements and infrastructure. Themes of gender and social inclusion were also identified as cutting across all these priority areas. Beyond the NAPA, the Nepalese government also developed a participatory process for setting up LAPAs. To date, 70 of these plans have been prepared in 14 districts of the country's mid- and far western regions.

In 2011 Nepal launched a national Climate Change Policy, the main goals of which are climate change mitigation and adaptation, improving livelihoods through low carbon socioeconomic development, and supporting commitments to national and international climate change agreements. The policy also has the stated aim of ensuring that 'at least 80 per cent of the total funds available for climate change activities flow to the grassroots level' (Government of Nepal 2011: 4).

Strategies for realising the goals of the Climate Change Policy have been identified, and the Nepalese government is in the process of implementing some of these, for example through its Low Carbon Economic Development Strategy. There is also a plan to use a climate-relevant budget code to track expenditure in the national budget. Several national mechanisms have been put in place to implement these strategies. Formed in 2009, the Climate Change Council is a 25-member high-level coordination body chaired by Nepal's prime minister. In the same year the Multi-stakeholder Climate Change Initiatives Coordination Committee was set up to help to implement collaborative programmes. Its members include representatives from relevant ministries and institutions, international and national NGOs, academia, the private sector and development partners. Reporting to and liaising with these two bodies, the Ministry of Science, Technology and Environment (MOSTE) is the lead ministry for implementing the UNFCCC's protocol and coordinating other climate change-related activities in Nepal.

Finance

Climate finance in Nepal comes from various sources, including bilateral partners and UNFCCC funds, and it is channelled both directly to the government and through intermediaries such as NGOs, UNDP and multilateral development banks. At the time of writing there were no plans for a national basket fund.

The government of Nepal has moved to incorporate climate change into national budget planning, however, with the implementation of a climate change budget code from the fiscal year 2011/12 onwards. This budget code enables climate change expenditure to be tracked; it emerged from a Climate Public Expenditure and Institutional Review conducted by the country's National Planning Commission in 2011.

Box 2.4 Main features of climate change planning in Nepal

Policies

- LAPAs
- Climate Change Policy
- Low Carbon Economic Development and Climate Resilience Strategy (under development)

Underlying political and climatic factors

- Adoption of a new constitution has slowed political processes
- High-level political engagement

Cross-sector mechanisms for LCRD

- Climate Change Council
- Multi-stakeholder Climate Change Initiatives Coordination Committee

Finance

- Channelled direct to government and through intermediaries such as NGOs, UNDP and multilateral development banks
- No national fund

Implementation

- Ministry of Science, Technology and Environment
- Partnerships with other ministries and programmes

Thus, we see that the four countries have all put in place a flagship policy explicitly addressing the climate agenda, developed cross-sectoral mechanisms and financing systems for implementation across government. The main differences between them are the focus of the policy document; the type of cross-sectoral mechanisms and extent of political support; and how the respective governments are seeking to channel their climate finance. We discuss these arrangements and their implications below and in more detail in the empirical chapters.

Conclusions

Drawing on the discussion of international frameworks, existing research, and these country case studies we identify a number of more general findings in three areas relating to LCRD planning in the LDCs.

Overarching drivers

The international negotiations have provided some initial drivers for LDCs to develop national frameworks such as NAPAs, NAMAs and NAPs and provided an initial framing for governments to engage with climate change as an international relations issue.

There are also a range of other national drivers to take forward the agenda such as access to finance, reducing the economic costs from vulnerability to climate change and reducing poverty, as well as providing the opportunity to leapfrog to new technologies.

Where countries have experienced extreme weather events and these have been linked to climate change, this has also created political impetus for change. Along with climate vulnerabilities identified through national planning processes, such events have formed the backdrop for increased efforts and political focus on LCRD. The case of Bangladesh, in particular – where cyclones and flooding have led to a focus on limiting damage to coastal infrastructure and people – shows that specific risks with high economic and social costs can help to create political momentum for a new form of climate change planning.

High-level political support has been particularly important in some cases. The climate change strategies in Ethiopia and Rwanda had early, high-level political support from their prime ministers and presidents, respectively. In both cases this gave the agenda a high profile and brought in high-level institutions and actors. This political momentum and leadership lent support to new policies.

National politics and drivers have also shaped the domestic climate agenda. Stakeholders in Bangladesh, particularly those from civil society organisations, have strongly expressed the view that adaptation should be prioritized in the country's climate change responses, and that low carbon measures are the responsibility of developed nations. A similar view emerged in Nepal, where the use of loans for a climate initiative (the PPCR) was deemed to be unacceptable by some actors due to the perception that adapting to climate change is the responsibility of developing countries and should therefore be addressed through grants. These opinions have influenced the climate change programmes and priorities in both countries.

Policy responses

International climate change policies provided the initial framework within which governments in the LDCs formulated their climate change responses, but since 2009 some have been developing strategies based more on their national development priorities. Several of these seek to address mitigation, adaptation and development priorities within one overarching framework.

Existing research on early climate programmes suggests that particular challenges for national LCRD policy responses may concern relationships within and between scales – with ownership being a particularly contested issue. Trade-offs and synergies also remain a challenging dimension of an LCRD agenda, with very few empirical data on where and how these might be found.

National policy responses have focused on particular aspects of the LCRD agenda. Initial policies in Bangladesh and Nepal, for example, prioritised adaptation but low carbon measures have since been incorporated as additional elements that might be considered with sufficient funding. The strategies in Ethiopia and Rwanda, however, established green growth and low carbon development upfront, and linked this to the idea of becoming policy leaders on the international stage, as well as to the potential for co-benefits in other areas such as energy access.

Governments are developing cross-sectoral institutional mechanisms to support ownership and address the policy coordination challenges of an LCRD agenda. All four of the focus countries have put institutional mechanisms in place to coordinate the climate change responses of different sectors such as high-level ministerial committees, multi-stakeholder committees and existing environmental committees. Some also have cells or focal points in other ministries to help to build ownership and to bring the agenda into the mainstream activities of key sectors.

Financial systems

Access, sustainability and scope are important issues for LDCs seeking to make use of international climate finance. The imbalance of adaptation and mitigation funding has been controversial, as have the differing modes of access for climate finance which LDCs can find cumbersome.

Setting up national climate funds allows governments to channel climate finance to agreed national priorities and maintain oversight over climate-relevant spending. Rwanda, Ethiopia and Bangladesh have all set up national basket funds to channel climate finance according to their priorities. These pool funds from a variety of sources for use in an overarching programme or policy, systematically addressing climate risks and opportunities. Nepal does not have such a fund and instead relies on separate actors funding specific activities. However, Nepal does plan to use a climate change budget code to track climate-related expenditure. Such arrangements give governments different levels of control over their climate change portfolios.

The national climate funds provide finance for both state and non-state actors – often line ministries, local governments and NGOs – to implement aspects of national policy and bring the climate change agenda into the mainstream. In Rwanda, where such mainstreaming is a key aim, the national climate fund offers a dedicated funding window for this purpose. Similarly, Ethiopia's fund has been set up to finance investments that deliver outcomes related to the country's CRGE strategy results areas. In Bangladesh, national funds – both donor-supported and government-owned – are used to support the priorities of the country's Climate Change Strategic Action Plan. These funds finance both adaptation and mitigation, often with strong development co-benefits, but there is no explicit intention to merge the agendas.

In conclusion, this chapter has shown that while international negotiations provide the context for national LCRD planning, national factors such as political will, existing priorities, climate risks, and national politics play key roles in how the policies and mechanisms are put in place for implementation.

References

African Development BankGroup (2013) *Rwanda energy sector review and action plan*. Available at http://www.afdb.org/fileadmin/uploads/afdb/Documents/Project-and-Operations/Rwanda_-_Energy_Sector_Review_and_Action_Plan.pdf (accessed 29 March 2016).

Alam, K., Shamsuddoha, M. D., Tanner, T., Sultana, M., Huq, M. J. and Kabir, S. S. (2011) The political economy of climate resilient development planning in Bangladesh. *IDS Bulletin*, 42: 52–61.

Ayers, J. and Dodman, D. (2010) Climate change adaptation and development I: the state of the debate. *Progress in Development Studies*, 10(2): 161–168.

Ayers, J., Anderson, S. and Kaur, N. (2011) Negotiating climate resilience in Nepal. *IDS Bulletin Special Issue: Political Economy of Climate Change*, 42: 70–79.

Bass, S., Wang, S., Ferede, T. and Fikreyesus, D. (2013) *Making growth green and inclusive: the case of Ethiopia*. Paris: OECD. Available at http://www.oecd-ilibrary.org/environment/making-growth-green-and-inclusive_5k46dbzhrkhl-en (accessed 27 March 2016).

Bird, N., Billet, S. and Colon, C. (2011) *Direct access to climate finance: experiences and lessons learned*. Discussion Paper. New York: UNDP/London: ODI. Available at http://www.odi.org/sites/odi.org.uk/files/odi-assets/publications-opinion-files/7479.pdf (accessed 24 March 2016).

Bulkeley, H. and Betsill, M. (2005) *Cities and climate change: urban sustainability and global environmental governance*. Abingdon: Routledge.

Ellis, K., Cambray, A. and Lemma, A. (2013) *Drivers and challenges for climate compatible development*. CDKN Working Paper. London: CDKN. Available at http://cdkn.org/wp-content/uploads/2013/02/CDKN_Working_Paper-Climate_Compatible_Development_final.pdf (accessed 29 March 2016).

Evans, A. (2012) *Resources, risk and resilience: scarcity and climate change in Ethiopia*. New York: Centre on International Cooperation, New York University.

Fikreyesus, D., Kaur, N., Kallore, M. and Ayalew, L. (2014) *Public policy responses for a Climate Resilient Green Economy in Ethiopia*. IIED Research Report. London: IIED. Available at http://pubs.iied.org/10066IIED.html (accessed 29 March 2016).

Fisher, S. (2013) *Low-carbon resilient development in the least developed countries*. IIED Issue Paper. London: IIED. Available at http://pubs.iied.org/17177IIED.html (accessed 16 December 2015).

Fisher, S. (2014) The emerging geographies of climate justice. *Geographical Journal*, 181(1): 73–82.

Fisher, S. and Mohun, R. (2015) *Low carbon resilient development and gender equality in the least developed countries*. IIED Issue Paper. London: IIED. Available at http://pubs.iied.org/10117IIED.html?b=d (accessed 16 December 2015).

Fisher, S. and Slaney, M. (2013) *The monitoring and evaluation of climate change adaptation in Nepal: a review of national systems*. IIED Research Report. London: IIED. Available at http://pubs.iied.org/10064IIED.html (accessed 23 March 2016).

Government of Ethiopia (2011) *Ethiopia's Climate Resilient Green Economy Strategy*. Addis Ababa: Ministry of Environment Protection and Forestry.

Government of Nepal (2010) *National Adaptation Programme of Action (NAPA) to climate change*. Kathmandu: Ministry of Science, Technology and Environment.

Government of Nepal (2011) *Climate change policy*. Kathmandu: Government of Nepal. Available at http://www.climatenepal.org.np/main/?p=research&sp=onlinelibrary&opt=detail&id=419 (accessed 7 January 2016).

Government of Rwanda (2011) *Green growth and climate resilience: national strategy for climate change and low carbon development*. Kigali: Government of Rwanda. Available at http://cdkn.org/wp-content/uploads/2010/12/Rwanda-Green-Growth-Strategy-FINAL1.pdf (accessed 29 March 2016).

Government of Rwanda (2012). *Operations manual: Environment and Climate Change Fund (FONERWA) Design Project*. Kigali: Government of Rwanda.

Government of Rwanda (2013) *Second Economic Development and Poverty Reduction Strategy (EDPRS2)*. Kigali: Government of Rwanda. Available at http://www.rdb.rw/uploads/tx_sbdownloader/EDPRS_2_Main_Document.pdf (accessed 20 March 2015).

Green Climate Fund (2015) *Green Climate Fund approves first 8 investments*. Incheon, Republic of Korea: Green Climate Fund. Available at http://www.greenclimate.fund/documents/20182/38417/Green_Climate_Fund_approves_first_8_investments.pdf/679227c6-c037-4b50-9636-fec1cd7e8588 (accessed 12 January 2016).

Haviland, C. (2015) *Why is Nepal's new constitution controversial?* London: BBC News. Available at http://www.bbc.co.uk/news/world-asia-34280015 (accessed 12 February 2016).

Jones, L. and Carabine, E. (2013) *Exploring the political and socio-economic drivers of transformational climate policy: early insights from the design of Ethiopia's Climate Resilient Green Economy strategy*. ODI Working Paper. London: ODI.

LDC Group Chair (2015) *LDC Group at UN climate change negotiations*. Press release. 30 November. Available at https://ldcclimate.wordpress.com/2015/11/30/press-release-climate-finance-pledges-clear-backlog-but-urgent-adaptation-funds-needed-for-2020-and-beyond/ (accessed 23 March 2016).

Lockwood, M. and Cameron, C. (2012) *Approaches to low carbon energy and development: bridging concepts and practice for low carbon climate resilient development*. Brighton: IDS-DFID Learning Hub.

Nakhooda, S. and Norman, M. (2014) *Climate finance: is it making a difference?* ODI Report. London: ODI. Available at http://www.odi.org/sites/odi.org.uk/files/odi-assets/publications-opinion-files/9359.pdf (accessed 22 February 2016).

Nash, E. and Ngabitsinze, J. (2014) *Low-carbon resilient development in Rwanda*. IIED Country Report. London: IIED. Available at http://pubs.iied.org/10065IIED.html (accessed 12 March 2016).

Newell, P., Phillips, J. and Pueyo, A. (2014) *The political economy of low carbon energy in Kenya*. IDS Working Paper No. 445. Brighton: IDS.

Pattberg, P. and Stripple, J. (2008) Beyond the public and private divide: remapping transnational climate governance in the 21st century. *International Environmental Agreements: Politics, Law and Economics*, 8(4): 367–388.

Pervin, M., Sultana, S., Phirum, A., Camara, I. F., Nzau, V. M., Phonnasane, V., Khounsy, P., Kaur, N. and Anderson, S. (2013) *A framework for mainstreaming climate resilience into development planning*. IIED Working Paper. London: IIED. Available at http://pubs.iied.org/10050IIED.html (accessed 16 December 2015).

Rai, N., Huq, S. and Huq, M. J. (2014) Climate resilient planning in Bangladesh: a review of progress and early experiences of moving from planning to implementation. *Development in Practice*, 24(4): 527–543.

Schipper, E. L. F. (2006) Conceptual history of adaptation in the UNFCCC process. *Review of European Community and International Environmental Law*, 15: 82–92.

Seballos, F. and Kreft, S. (2011) Towards an understanding of the political economy of the PPCR. *IDS Bulletin*, 42(3): 33–41.

Stockholm Environment Institute (SEI) (2009) *Economics of climate change in Rwanda*. Stockholm: SEI.

Shankland A. and Chambote R. (2011) Prioritising PPCR investments in Mozambique: the politics of 'country ownership' and 'stakeholder participation'. *IDS Bulletin Special Issue: Political Economy of Climate Change*, 42: 62–69.

Suckall, N., Stringer, L. and Tompkins, E. (2014) Presenting triple-wins? Assessing projects that deliver adaptation, mitigation and development co-benefits in rural sub-Saharan Africa. *AMBIO*, 44: 34–41.

Tanner, T. and Allouche, J. (2011) Towards a new political economy of climate change and development. *IDS Bulletin*, 42: 1–14.

Tanner, T., Mensah, A., Lawson, E., Gordon, C., Godfrey-Wood, R. and Cannon, T. (2014) *Political economy of climate compatible development: artisanal fisheries and climate change in Ghana*. IDS Working Paper No. 446. Brighton: IDS.

Tenzing, J., Gaspar-Martins, G. and Jallow, B. P. (2015) *LDC perspectives on the future of the Least Developed Countries Fund*. LDC Paper Series. London: CDKN. Available at https://ldcclimate.files.wordpress.com/2012/05/ldc-perspectives-on-the-ldcf.pdf (accessed 30 March 2016).

Tompkins, E., Mensah, A., King, L., Tran, K., Lawson, E., Hutton, C., Hoang, V., Gordon, C., Fish, M., Dyer, J. and Bood, N. (2013) *An investigation of the evidence of benefits from climate compatible development*. Sustainability Research Institute Paper No. 44. Leeds: University of Leeds.

UNFCCC (2011) *Cancun adaptation framework*. Available at http://unfccc.int/adaptation/items/5852.php (accessed 16 December 2015).

Uprety, B. (2015) *Financing climate change adaptation in LDCs*. Available at http://www.iied.org/financing-climate-change-adaptation-ldcs (accessed 22 March 2016).

3 Storylines in national climate planning and politics

Finance, ownership and shifting synergies

Susannah Fisher, Misgana E. Kallore, Nazria Islam, Lidya Tesfaye and John Rwirahira

Introduction

A particular challenge in the emerging discourse of LCRD is how to bring together the agendas of mitigation, adaptation and development, if at all. Within global climate discourse, mitigation (reducing greenhouse gas emissions) has been traditionally the responsibility of developed countries, while developing countries have been supported to focus on adapting to the potential effects of climate change and maintaining their development trajectory. However, there are theoretical reasons for bringing these policy areas together, such as leveraging better outcomes in each policy area, maximising efficiency in using limited resources and reducing the possibility that one area may impact negatively on another (i.e. avoiding or managing trade-offs). One part of the agenda (for example adaptation) can also be used to help to incentivise action in another that might be less of a priority, such as mitigation (Ayers and Huq 2009). However, this is an emerging area of planning and there is little evidence of how such benefits might be realised in LDCs, and how best to plan for them.

In this chapter we examine the agenda-setting processes involved in creating the national climate change plans and strategies of three countries that seek (in their rhetoric, at least) to bring together the mitigation, adaptation and development agendas at national scale. We ask how and why countries are adopting this type of LCRD approach, and how storylines and discourse coalitions shape these processes. The chapter is based on research conducted for an IIED research project on drivers for LCRD (Fisher et al. 2014; our methods are described in more detail in the next section).

In this chapter we argue that:

- Access to climate finance, demonstrating global leadership and supporting existing national priorities were key storylines about initiating an LCRD policy for stakeholders in Bangladesh, Ethiopia and Rwanda that were supported by related incentives.
- Maintaining national ownership is an important aspect of building on the initial plan.

- Existing national storylines can shape priorities within an LCRD agenda.
- Stakeholder storylines about synergies within an LCRD policy approach are disparate, and no clear coalitions have as yet emerged among stakeholders in either Ethiopia or Rwanda. In Bangladesh two policy communities have coalesced around two separate approaches. The storylines that have become institutionalised within policy are also partly dispersed.
- Questions relating to potential trade-offs can be marginalised, excluded from agenda-setting discussions, and need to be considered in more detail when moving towards implementation.

We start the chapter with a brief discussion of the theory of how adaptation, mitigation and development might be brought together. We then outline the approach to LCRD and the storylines about initiating an LCRD agenda in each country. We examine in detail stakeholder storylines and their institutionalisation into policy, looking specifically at areas of synergy within an LCRD agenda. We conclude with a discussion of our findings and their implications.

Bringing together the agendas: theory and methods

As discussed in Chapter 2, there is a growing discourse concerned with bringing adaptation, mitigation and development together within one agenda, but very little empirical work has been done in developing countries – especially LDCs – on whether or how this works in practice; there are also theoretical challenges involved. Klein et al. (2007) highlight the different temporal and spatial scales at which mitigation and adaptation are effective, with mitigation offering medium- to long-term benefits at global scale and adaptation often having more immediate and local benefits. They also suggest that it is difficult to compare the costs and benefits of adaptation and mitigation policies; the latter are more easily quantified, whereas adaptation is notoriously difficult to measure (see also Brooks et al. 2013).

In terms of institutional approaches and mechanisms, different options are available to national governments in LDCs. Drawing on a review of all LCRD policies in LDCs from previous work (Fisher 2013), we outline three main approaches to finding potential synergies between the agendas:

- A single policy, for example a national campaign to distribute solar lanterns as a decentralised renewable energy solution. This type of campaign or policy could offer benefits in all three areas: mitigation through renewable technology, adaptation through addressing the underlying causes of vulnerability and development through better educational outcomes and income diversification (by being able to work in the evenings).
- An overarching policy objective within which specific policies address different strands of the agenda; these policies could focus on mitigation, adaptation or development.

- Implementing multiple objectives simultaneously with a single funding mechanism. This does not necessarily imply any synergies in implementation beyond a general political will to support both agendas. Policies may also be designed primarily to address one agenda – mitigation, adaptation or development – while at the same time making some contribution to another.

There are local examples of using approaches like these to leverage co-benefits or triple wins as we saw in Chapter 2, but the question is how commonly such outcomes can be realised and how significant their benefits are at national scale. A political economy analysis allows us to consider why particular approaches have been chosen, what the ideas behind them are and whether any key dimensions have been marginalised in existing national approaches. It also allows us to consider how such policies progress from planning to implementation.

This chapter is based on semi-structured interviews conducted with key stakeholders involved in the agenda-setting and formulation stages of the LCRD policy processes in the three focus countries. Between 20 and 25 interviews were conducted in each country, across four stakeholder groups: government officials; research and policy institutes; international organisations and development partners; and civil society and the private sector. Questions were asked about the process through which plans were developed and about stakeholders' views on some key factors in LCRD planning, such as their motivations and what the terms 'low carbon' and 'resilience' meant to them in different contexts. We also undertook a detailed policy and institutional analysis of how adaptation, mitigation and development were treated and managed. Finally, for Ethiopia and Bangladesh we conducted an online survey of the information that stakeholders used to plan for LCRD and how they shared information and worked with each other during the planning phase (we were unable to conduct the same survey for Rwanda due to the timing of the research).

The interviews and surveys were conducted with a sample of the policy-making community available for interview between July 2013 and February 2014, identified through purposive sampling to cover the key institutions and actors. Therefore the analysis covers a range of ideas but can neither claim to represent all views, nor can it indicate how dominant each view is. The interviews and surveys refer to policy processes that have been unfolding for several years; it is likely that during this time views and ideas have shifted. This is also a fast-evolving area of policy and new developments have been announced since the interviews and surveys took place. We have addressed this with questions on particular processes, but our analysis focuses on storylines at a particular time in 2013–14.

This chapter focuses on storylines and discourse coalitions, using the framework discussed in Chapter 1 and Hajer's (1995) concept of 'storylines'. A storyline is the shorthand way in which actors understand and describe issues, based on a particular set of assumptions and value judgements. Discourse coalitions are groups of actors who subscribe to the same storyline, though they may not have the same rationale for this or feel part of a distinct 'group'.

Particular storylines and coalitions gain traction when they become embedded within institutions and policies. This is the result of an interaction between the discursive power of the storylines themselves and other factors such as material power and resources, supporting incentives and institutional structures. We do not suggest that this reflects linear causation – that storylines are translated directly into institutions and policies – rather it is an iterative process, overlapping and fragmented, in which one reinforces the other. Stakeholder storylines and their institutionalisation emerge together, with institutionalisation occurring where a storyline is given sufficient weight, in context.

Moving on to the empirical section of this chapter, we first examine the reasoning behind governments' initial engagement with an LCRD policy process, before going on to look in more detail at potential synergies within an LCRD approach.

Initiating an LCRD agenda

This section examines why, in the face of many pressing development concerns, some LDC governments have initiated an LCRD agenda.

Bangladesh, Ethiopia and Rwanda developed their national plans and strategies between 2009 and 2011; all include aspects of adaptation, mitigation and development. The main 'flagship' policies are summarised in Table 3.1. As this shows, all three governments have adopted an overarching policy objective, bringing together all elements of the climate change and

Table 3.1 Flagship LCRD plans in Bangladesh, Ethiopia and Rwanda

Country	Plan	Approach to LCRD	Mechanisms
Bangladesh	Bangladesh Climate Change Strategic Action Plan (2009)	Adaptation and mitigation separate pillars under one plan	Overarching policy objective Single funding mechanism
Ethiopia	Climate Resilient Green Economy Vision and Strategies (2010/11)	Initially, green growth and resilience as separate issues, now looking to mainstream both	Overarching policy objective Single funding mechanism
Rwanda	National Low Carbon and Climate Resilient Development (2010/11)	Mainstreaming into development	Overarching policy objective Single funding mechanism

development agendas, with financial mechanisms supporting both adaptation and mitigation. (See Chapter 2 for more detail on the policies, institutional frameworks and financial mechanisms in each country.)

Table 3.2 summarises the main policy storylines supporting national engagement with an LCRD agenda, based on storyline analysis of interview data and policies.

As the table shows, there is consensus among stakeholders in all the countries on three main storylines supporting the initial adoption of an LCRD approach: the opportunity to access climate finance; showing global leadership; and supporting existing national priorities for socioeconomic development. These are linked to related incentives; for example, the storyline about accessing climate finance is linked to economic incentives, while ideas about global leadership are related to reputational incentives. (The storylines specific to each country are discussed in more detail later in the chapter.)

Second, in addition to the storylines that were common to all three countries, we identified a range of national storylines that motivated stakeholders to take on an LCRD agenda and also, we suggest, supported the prioritisation of particular dimensions of their national plans. For example, stakeholders in Ethiopia talked about using LCRD as a means of boosting their development agenda. This storyline characterises LCRD as a means of supporting more ambitious policies, and ties in with the country's plans for green growth and greenhouse gas abatement, as well as a high level of political commitment to the Green Economy Strategy. In addition to this policy incentive, a reputational incentive is also operating here, reflecting a belief that successful implementation may mark the country out as a leader and earn it a role as a global player on the international stage.

Within the storylines identified in Bangladesh the overriding priority was adaptation, with additional specific concerns about national agenda ownership,

Table 3.2 Storylines supporting LCRD in Bangladesh, Ethiopia and Rwanda

Country	National storylines supporting LCRD
Bangladesh	Supporting existing national priorities Leadership role Low carbon must be funded through other means Adaptation as a priority Access to finance
Ethiopia	Supporting existing national priorities More ambitious policies, green growth Leadership role, becoming a global player Access to finance
Rwanda	Supporting existing national priorities Leadership role Own national path, aimed at achieving broader environmental sustainability Access to finance

energy security and effective targeting of the poorest sections of society. Some stakeholders in Bangladesh also felt strongly that mitigation efforts should be funded from external rather than domestic sources. They supported an LCRD agenda that reflected common but differentiated responsibility and prioritised adaptation measures.

In addition to the incentive of climate finance, Rwandan stakeholders considered that policies for low carbon resilience were closely linked to a broader domestic agenda aimed at achieving environmental sustainability; that is, to a particular vision for the country's future. This reflects a policy incentive, and is also linked to knowledge incentives around the broader environmental agenda. Rwanda's national plan and financing structure also embodies this broader approach to climate and the environment.

Finally, a sense of national ownership of climate actions was identified as important in all three countries, specifically in relation to taking the agenda forward following its initial adoption. There was a variety of storylines about national ownership in Bangladesh. Some stakeholders from government, development partners and civil society coalesced around a storyline of national leadership, stating that because Bangladesh's contribution to global greenhouse gas emissions is small there is little external pressure on it to undertake mitigation actions. Instead, issues such as energy scarcity are the main drivers for national LCRD. These priorities are reflected in national policies, the allocation of government resources to climate change programmes and the structure of the institutions set up to coordinate them. Consistent with this sense of national ownership, Bangladesh has placed great emphasis on being a leader among LDCs and on an inclusive approach to climate planning. As one stakeholder explained:

> The BCCSAP came about through a consultative process – not just by three or four consultants sitting in Dhaka. That is not correct. The BCCSAP document is well acclaimed among all the LDCs. Among the LDCs Bangladesh is placed number one for having this unique document. It was showcased at top meetings and negotiations. Even I had this experience. When I joined the Ministry I saw the arrival of a 15-member Nepalese delegation sent to discover how the Plan was formulated. A a two-day session was held at the Ministry and those who were associated with this preparation process delivered lectures here. I think that the BCCSAP is very important, and everyone knows about it.
> (Official at the MOEF 2013)

However, alternative views were held by some stakeholders from development agencies, government and civil society: some felt that there was a lack of political leadership in relation to low carbon pathways, while others felt that there was a need for wider engagement with the agenda across society.

In Ethiopia and Rwanda high-level political figures supported the LCRD approach and gave it initial impetus. However, according to interviews with civil society actors and others, national involvement in the LCRD policy process was also a key issue in Ethiopia, as the planning process unfolded. For example, stakeholders – even some involved in technical sub-committees – talked about struggling to understand what the country's CRGE work was about, as well as observing that only a few stakeholders were involved at key moments of the process. Bass et al. (2013) note the key technical role that international organisations such as the Global Green Growth Institute (GGGI) played in the CRGE process, and the importance of striking a balance between such international support such as that from the GGGI and building domestic engagement as noted by the stakeholder interviews.

Shaping areas of overlap: storylines and institutionalisation

We now go on to identify and describe the storylines about areas of synergy within LCRD in each national context. We then analyse which stakeholders support each storyline, and how the storylines have been institutionalised in the form of policies. We show that stakeholder storylines are largely dispersed about how LCRD policy areas might come together, and this dispersal is reflected to some extent within policy documents.

Table 3.3 summarises the main stakeholder storylines relating to LCRD in the three focus countries, and is based on analysis of interview data. Each set of storylines has been developed by identifying areas of commonality and overlap within the spectrum of viewpoints and actors in each country. They reflect the different values and beliefs among those involved about how a national LCRD agenda might be put into practice, as well as their assumptions about who is responsible for acting in response to climate change, who might benefit from LCRD actions, and who has the right to define the main issues and participate in the policy process.

It is evident that within each country the concept of LCRD has a variety of different meanings for different actors. There is also a diversity of views on how synergies might be found between the mitigation, adaptation and development agendas. The table shows this range of viewpoints, from treating the agendas as separate (at the top) to integrating them entirely (at the foot). Looking at individual storylines in more detail, the first, 'no need for low carbon', reflects an idea of 'common but differentiated responsibilities' and the view that developing countries should not be held responsible for other (developed) countries' historic carbon emissions. It involves rejecting the idea of an intersection between adaptation and mitigation actions, as this muddies the waters on the issue of responsibility. This view is shared in particular by a bloc of countries mobilised by Southern solidarity, and is one that has been strongly expressed in international negotiations.

Moving down the table, the second storyline, 'separation', puts less emphasis on the issue of responsibility but maintains a separation between the adaptation and mitigation agendas. The third storyline, 'sequential', adds the dimension of time: adaptation is the priority, but low carbon measures

Table 3.3 Understandings of LCRD in Ethiopia, Bangladesh and Rwanda: storylines and coalitions

Storyline	Meaning	Stakeholders coalescing behind storylines		
		Bangladesh	Ethiopia	Rwanda
No need for low carbon	Adaptation is a national priority, mitigation is not a national issue	Some political figures Civil society		
Separation	Adaptation and mitigation are separate issues, with no linkages	Civil society Development partners	Some government ministries	Some government ministries and agencies Private sector
Sequential	Adaptation continues to be a priority, but low carbon growth should be included sequentially, over time	Government bodies linked to the environmental sector	Some development partners	Some government ministries and agencies Private sector
Alignment	Pursuing adaptation and mitigation agendas in parallel, without necessarily addressing overlaps or synergies	Government bodies linked to the environmental sector Research organisations and think tanks	Some government ministries	
Co-benefits	Prioritise those actions that have common mitigation and adaptation benefits (co-benefits), without curtailing development	Government bodies linked to the environmental sector Research organisations and think tanks	Some government ministries and agencies International organisations	
Complementarity	Agendas indirectly support each other		Some government ministries, research institutions and development partners	Government ministries and agencies International organisations Development partners

Storyline	Meaning	Stakeholders coalescing behind storylines		
		Bangladesh	Ethiopia	Rwanda
Complexity	The different agendas are complex and cannot be clearly distinguished, so are implemented together		Some government ministries and agencies	Government ministries and agencies International organisations Development partners
Low carbon as a feasible approach	Low carbon is increasingly a priority and more feasible than building resilience	Ministries dealing with power generation and energy		Government ministries and agencies International organisations Development partners
Integration and leveraging	Integration of adaptation and mitigation agendas; can result in better outcomes	Development partners	Some government ministries	Government ministries and agencies International organisations Development partners
Long-term sustainability	Implementing the low carbon and resilience agendas together leads to mutual socioeconomic and environmental benefits		Some government ministries	Government ministries and agencies International organisations Development partners

Source: Fisher et al. (2014), available at pubs.iied.org/10099IIED, © IIED.

may follow. 'Alignment' introduces the idea that the agendas should be implemented alongside one another, while 'co-benefits' goes further, suggesting that interventions that realise synergies between the climate agendas should be prioritised. 'Complementarity' reflects the notion that the agendas should indirectly support each other.

The 'complexity' and 'low carbon as a feasible approach' storylines reflect viewpoints that are based on pragmatism. The former casts the agendas as complex, difficult to define or separate, and so suggests that there is no real purpose to governments categorising policies or their potential benefits as falling under particular headings. The latter reflects the view that low carbon measures are relatively easy to identify and implement, and more practicable than measures aimed at building resilience, which should therefore be regarded as longer-term goals.

'Integration and leveraging', meanwhile, is a step on from the co-benefits storyline, the idea being to integrate and mainstream all actions into national policy to leverage better outcomes across all domains. The final storyline, 'long-term sustainability' is a broader framing, within which all agendas contribute to an overarching goal, and the details of how and why they might combine are less important.

Storylines in Bangladesh

In Bangladesh, several conflicting storylines are apparent. The views expressed by government stakeholders tended to coalesce around two broad storylines, indicating two distinct coalitions. On the one hand, those from government bodies focusing on environmental issues generally held on to a sequencing storyline, whereby adaptation measures would be prioritised but mitigation actions added over time. Some took the broadly similar view that actions giving rise to both mitigation and adaptation should be prioritised – the 'co-benefits' storyline – in order to meet international climate commitments; this perspective was also shared by some research and think tank experts who had been involved in designing Bangladesh's climate change plans. These storylines emphasise climate resilience and poverty reduction as overwhelming priorities.

On the other hand, those from government bodies focusing on energy – those typically working more closely with the private sector and development partners – saw low carbon development as a priority and as being more feasible than resilience measures. According to one official,

> The mitigation measures are comparatively easier to accommodate than the adaptation measures. For example, we have had to resettle people living along the coastline because they are affected by cyclonic events. They have a particular lifestyle and livelihood and when we resettle them it takes time for them to adapt to their new situations and for their new environments to accommodate them … It is this adaptation stage

for which long-term planning is needed and it takes time to adjust to all these things. It is a very difficult task. Compared to adaptation, mitigation is a little easier. For example, as part of our mitigation efforts we replaced the diesel sets with the solar irrigation sets.

Official at the Ministry of Power (2013)

There was some consensus among international organisations and development partners, who saw low carbon and resilience as separate from each other, with both requiring dedicated policy and actions. Some development partners, however, saw the need to integrate the two agendas into policy and actions. The 'no need for low carbon' viewpoint is held by groups of civil society actors and some political figures.

In terms of the storylines institutionalised in policy, the flagship BCCSAP recognises the need for adaptation and mitigation and emphasises a low carbon development pathway, and can be seen as representative of a sequential storyline. A few of Bangladesh's key LCRD policies also reflect a co-benefits storyline. The Climate Management Plan for the Agricultural Sector (Programme Support Unit 2009) was the first policy document to explicitly mention the potential for pursuing synergies between mitigation and adaptation measures. This has also been translated into practical actions in later documents, for example in the National Sustainable Development Strategy, which states that the country's 'afforestation program should be strengthened to take advantage of its effect on disaster risk reduction and climate change mitigation' (Government of Bangladesh 2013: 127).

Finally, despite storylines about adaptation being a particular priority, the 'low carbon as a feasible option' storyline also carries power within funding and policy circles. Although low carbon development goes largely unmentioned in policy documentation, Bangladesh has implemented many projects aimed at energy security, contributing to mitigation. Rai et al. (2014) have shown that around 40 per cent of national climate funds from the BCCRF are going into mitigation and low carbon projects including, for example, programmes to reduce system losses from energy distribution.

Storylines in Ethiopia

At the time of our research there were still many different views among state and non-state actors in Ethiopia as to what the country's CRGE initiative actually means. Our analysis highlights the breadth of storylines among stakeholders about how climate resilience and green economy agendas might be brought together, representing a spectrum of views on the degree of engagement between the agendas. For example, reflecting the 'complexity' storyline, an official at the Ministry of Agriculture (2013) said:

> There is no clear distinction between climate resilience and green economy that we can use to help to identify and apply to separate activities. They

cannot be addressed sequentially – i.e. one after the other – but need to be looked at simultaneously to support sustainable outcome[s].

Stakeholders also considered leveraging better outcomes as a potential benefit of LCRD, suggesting some support for a more transformative agenda.

The storylines and coalitions envisage finding varying synergies between adaptation, mitigation and development agendas. In general:

- Government stakeholders had a range of views on the issue, with no clear consensus around what this means in practice.
- The greatest consensus was among international organisations and development partners, who all saw a need for a complementary and leveraging approach.
- Private sector stakeholders understood the issue as part of corporate social responsibility, a perspective not shared by others.
- The research organisations had greater technical understanding, based on differences of scale in the agendas.

Turning to the storylines that were institutionalised in Ethiopia, the national Green Economy Strategy (which, along with the parallel Climate Resilient Strategy, supports the country's overall CRGE vision) does not explicitly consider how to improve climate resilience, nor does it consider climate change vulnerability as a specific criterion for prioritising interventions. However, it does consider the potential for making contributions to the targets of Ethiopia's national development plan as one such criterion, and this could be regarded as a way of taking resilience considerations into account. This indicates the use of either an 'alignment' or a 'co-benefits' storyline, to align green growth benefits with national priorities, or to ensure that areas of co-benefit are prioritised under both green growth and development plans.

The Climate Resilience Strategy for agriculture also used contributions to the national development plan as a criterion for prioritisation (OECD 2014). New, specific mechanisms and guidelines are being developed, however, that will offer more explicit guidance on merging the two climate agendas. For example, a Sectoral Reduction Mechanism is being planned to provide technical and financial coordination of plans (Government of Ethiopia 2014). In addition, a recently introduced CRGE strategy results matrix includes co-benefits indicators for five policy areas: food security; rural incomes and green jobs; health and well-being; access to basic services; and gender and differential vulnerability.

Together these developments suggest high-level commitment to moving towards mainstreaming climate change policies within national development planning and realising climate co-benefits. The national development plan clearly outlines building a climate resilient green economy as key to long-term sustainability in the context of climate change, thus aligning with national priorities. The co-benefits narrative also comes through strongly in the CRGE results matrix, although the detail defining what resilient, green industry actually is and how CRGE activities will support the five areas of co-benefits has yet to materialise.

Storylines in Rwanda

Many stakeholders in Rwanda, regardless of the storyline they emphasised, noted the close relationship with environmental issues and sustainability; this is what they considered a transformative, low carbon resilient agenda to look like. For the majority, bringing together climate resilient and low carbon development also meant working to increase the awareness and understanding of stakeholders involved in planning and implementing the country's priorities. As Table 3.2 shows, storylines were spread across a spectrum, with some stakeholders suggesting that there is no need to address synergies while others considered it important to do so.

In Rwanda, policies and institutions tend to adhere to an overarching storyline concerned with long-term environmental sustainability, rather than any more specific storylines dealing with the relationship between agendas. The purpose of the country's NSCCLCD is to guide national planning in an integrated way, mainstream climate change measures within all sectors of the economy, and position Rwanda to access international LCRD funding.

The NSCCLCD informed the development of the latest iteration of Rwanda's key five-year development plan, the EDPRS 2, published in 2013. This document frames climate change as an opportunity 'to ensure sustainability of interventions through environment mainstreaming, eco-system protection and rehabilitation as well as tapping into the growing international pool of green investments' (Government of Rwanda 2013: 13). The influence of the NSCCLCD is evident from the inclusion in the EDPRS 2 of mainstreaming strategies as part of the development planning framework, particularly in relation to integrated land-use planning and management, sustainable small-scale energy installations in rural areas and low carbon urban systems (Nash and Ngabitsinze 2014). It also identifies green growth as a priority area for achieving economic transformation, and includes a results matrix with objectives for green urbanisation and environmental standards.

Discussion of synergies

Based on this analysis, we suggest that two broad findings can be drawn from our examination of storylines. First, there are multiple stakeholder storylines about synergies within LCRD and stakeholders' views are diffuse, covering a range of perspectives. Individuals within the same organisations sometimes coalesce around different storylines, showing the importance of personal reflection as well as institutional factors in the way in which stakeholders approach the agenda. While some stakeholder storylines are institutionalised, and therefore influence policies and financing decisions, others remain confined to stakeholder discourse.

The storylines institutionalised into policies represent a smaller subset of approaches than those within stakeholder storylines but still retain some diversity at national level. Based on our analysis, it is notable that the more transformative storylines are not among those institutionalised in the case study countries. In Ethiopia, for instance, this applies to storylines based on integrating agendas in order to leverage better outcomes, while in Bangladesh storylines rejecting mitigation actions over the issue of responsibility have remained on the margins, eclipsed by policy priorities such as financing renewable energy projects.

This is important for several reasons. In the absence of a coherent vision it will be more difficult to secure multi-sectoral synergies involving multiple actors. In the context of climate change there is also the risk that mitigation and adaptation measures may conflict in the medium to long term if both agendas are not taken into account in some systematic way. That is, without planning, adaptation measures might result in higher greenhouse gas emissions and mitigation measures might increase countries' exposure to particular climate risks. Some common understanding needs to be reached in this area, when moving towards implementation. In addition, although ideas may become more defined as the plans move to implementation and stakeholders are forced to confront the detail of the policies, the storylines and coalitions developed at the agenda-setting phase may continue to shape the questions asked and the approach taken. This opens up the possibility of multiple approaches to implementation, with individuals coalescing around different perspectives.

Our second main finding relates to storylines that are omitted or marginalised in the agenda-setting phase. These are instances where issues seem to demand a policy position but no such positions are made explicit, either within stakeholder discourse or in policies. For example, the issue of trade-offs – the potential impact of one agenda on another that need to be balanced or managed – did not emerge as a major consideration among the stakeholders we interviewed, and rarely featured in the policy frameworks of the case study countries. We must recognise that this could be due to our methodology and sampling, but while stakeholders talked about policies that might imply trade-offs – between energy choices, use of fertilisers and modifying livelihoods, for instance – they did not seem to conceptualise these as leading to potential trade-offs in dimensions such as growth or development. This finding is supported by the analysis of Bass et al., who note that in Ethiopia 'some of the CRGE's plans – notably to shift beef producers to poultry production – are extremely challenging. They might yield major GHG [greenhouse gas] reductions, but at the cost of considerable social upheaval, as well as removing what can be the ecologically optimal use of rangelands' (2013: 19).

A few stakeholders did note potential trade-offs, however. An official from Bangladesh's MOEF (2013), for example, described trade-offs around cost and choosing renewable energy sources at the individual level, saying that 'in the case of the Solar Home Systems, the solar panels ... are [expensive]. For a poor man in a rural household, [they] cannot afford to go for Solar Home Systems or panels. They are very costly. Technically also these are not adequately efficient. It is an emerging area, unfolding area. So unless price-wise it is affordable, it is difficult. Affordability is an issue.' Such issues were not widely addressed in interviews, however.

In other situations, the potential synergies within an LCRD agenda were framed so broadly that the detail of any trade-offs or how conflicts might play out in implementation were left to one side, at least in the agenda-setting phase. For example, in Rwanda the overarching storyline of environmental sustainability does not tackle the detail of how options for mitigation or adaptation might be prioritised, and therefore what trade-offs there might be. These details may emerge when moving to implementation or be explored on a case-by-case basis as and when the issue of trade-offs arises, but when they do it will be important to ensure that stakeholder storylines about synergies are re-examined to take into account these aspects of the debate. The nature of policy storylines as understood by Hajer (1995) is that they become 'shorthand' and can therefore close off questions outside the framing. Stakeholders will need to be explicitly aware of this dimension to ensure that such questions are addressed. As discussed earlier, existing national development priorities can shape agendas and these often contain commitments to progressive and equitable development outcomes. However, it is important to keep a critical eye on the new dimensions of equity that an LCRD agenda throws up, particularly where they relate to impacts on vulnerable communities, such as trade-offs between long-term goals, short-term benefits and poverty alleviation.

We have argued elsewhere that some dominant storylines can eclipse others, masking them with apparent consensus (Fisher 2012). Although further research is needed, there is some evidence of this here, raising the possibility that overarching discourses about win-win benefits, finance and readiness are effectively closing off or marginalising questions about exactly how agendas should come together, potential trade-offs and the implications of a transformative agenda for an economy and society.

Conclusions

In summary, we have shown, first, that the incentives of access to international climate finance, the opportunity to pursue existing national priorities, and global leadership and a sense of national ownership are common across countries pursuing an LCRD agenda. We have also found that existing national storylines can shape priorities within an LCRD agenda. Second, stakeholder storylines about the agenda are disparate and no single coherent coalition has emerged to take a particular approach forward. Often multiple storylines are institutionalised, becoming embedded in policy and financing decisions, resulting in a potential lack of clarity and direction in delivery. Third, some key storylines – for example about trade-offs – may have been omitted or marginalised in policy discourse in the agenda-setting phase. While this is understandable, it will be important to raise such questions as plans move towards implementation, and detailed plans are agreed.

These findings have implications for policy and practice. First, the existence of a national flagship climate plan is regarded as a crucial part of a climate response both in the academic literature on international political economy and in climate programming and evaluation processes. One of the major climate programmes, the PPCR, for instance, uses the existence of such a plan as an indicator of the extent to which countries have achieved climate change mainstreaming. Similarly, work on evaluating the effectiveness of adaptation measures has included institutional capacity as a criterion, of which one aspect is a national plan (see Brooks et al. 2013). However, our analysis shows that while such a plan is an important first step it does not necessarily mean that there is consensus or clarity about the path ahead. All three of our case study countries have plans in place, but despite this there is still a large breadth of discourses, and hence a range of potential paths and approaches that could emerge.

Further critical reflection on this point is therefore needed: it may be a particular characteristic of governments or national planning processes in LDCs, but international agreements rely on national implementation. The existence of a multiplicity of approaches and visions even after flagship national policies have been established is therefore a potential cause for concern. This is particularly true in view of the general shift towards mainstreaming (that is, towards integrating climate change concerns into existing sectors rather than relying on a stand-alone plan). Mainstreaming involves a much wider range of stakeholders in the policy process and so would require a broader coalition to support the chosen approach, in order to secure cross-sectoral synergies and avoid maladaptation.

Second, this chapter also raises questions about the process and time needed to build a national LCRD agenda. Our findings suggest that developing an LCRD approach is an ongoing process that needs to go further than institutional plans, mechanisms and systems, to develop a truly coherent vision of a transition pathway and what it means across government and society. Readiness for finance is not just about a country having adequate financial systems in place, it is also about whether it is in a position to use the finance it receives to realise transformative, cross-cutting objectives in the context of widespread debate and consensus around the chosen approach.

To conclude, this chapter has shown that LCRD is being renegotiated and translated within national planning processes in the LDCs to match national contexts, politics and priorities. In order better to understand these processes, future work needs to provide further empirical evidence, and to use this evidence to inform theorising about climate politics at multiple scales. It should look beyond the flagship national climate policies to explore the underlying processes of consensus and contention which will shape the LCRD agenda as it moves towards implementation.

References

Ayers, J. and Huq, S. (2009) The value of linking mitigation and adaptation: a case study of Bangladesh. *Environmental Management*, 43(5): 753–764.

Bass, S., Wang, S., Ferede, T., and Fikreyesus, D. (2013) *Making growth green and inclusive: the case of Ethiopia*. Paris: OECD. Available at http://www.oecd-ilibrary. org/environment/making-growth-green-and-inclusive_5k46dbzhrkhl-en (accessed 27 March 2016).

Brooks, N., Anderson, S., Burton, I., Fisher, S., Rai, N. and Tellam, I. (2013) *An operational framework for Tracking Adaptation and Measuring Development (TAMD)*. IIED Climate Change Working Paper No. 5. London: IIED. Available at http://pubs.iied.org/10038IIED.html (accessed 27 March 2016).

Programme Support Unit (of the Agricultural Sector Programme Support, Phase II) (2009) *Climate Management Plan: Agricultural Sector Bangladesh*. Available at http://www.bdresearch.org.bd/home/climate_knowledge/cd1/pdf/Bangladesh%20and%20climate%20change/Disaster%20management,adaptation/FINALClimateManagementPlan_Agriculture.pdf (accessed 20 March 2016).

Fisher, S. (2012) Policy storylines in Indian climate politics: opening new political spaces? *Environment and Planning C: Government and Policy*, 30(1): 109–127.

Fisher, S. (2013) *Low-carbon resilient development in the least developed countries: emerging issues and areas of research*. IIED Issue Paper. London: IIED. Available at http://pubs.iied.org/10049IIED.html (accessed 27 March 2016).

Fisher, S., Fikreyesus, D., Islam, N., Kallore, M., Kaur, N., Shamsuddoha, Md., Nash, E., Rai, N., Tesfaye, L. and Rwirahira, J. (2014) *Bringing together the low-carbon and resilience agendas: Bangladesh, Ethiopia, Rwanda*. IIED Working Paper. London: IIED. Available at http://pubs.iied.org/10099IIED.html (accessed 27 March 2016).

Government of Bangladesh (2013) *National Strategy for Sustainable Development*. Available at http://www.plancomm.gov.bd/national-sustainable-development-strategy/ (accessed 18 September 2014).

Government of Ethiopia (2014) *Climate Resilient Green Economy (CRGE) Facility: operations manual*. Draft. Addis Ababa: Government of the Federal Democratic Republic of Ethiopia. Available at http://www.ethcrge.info/resources/CRGE%20Facility%20Operation%20Manual_all%20sections.pdf (accessed 20 March 2016).

Government of Rwanda (2011) *Green growth and climate resilience: National Strategy for Climate Change and Low Carbon Development*. Kigali: Government of Rwanda. Available at http://cdkn.org/wp-content/uploads/2010/12/Rwanda-Green-Growth-Strategy-FINAL1.pdf (accessed 27 March 2016).

Government of Rwanda (2013) *Economic Development and Poverty Reduction Strategy II*. Kigali: Government of Rwanda. Available at http://www.rdb.rw/uploads/tx_sbdownloader/EDPRS_2_Main_Document.pdf (accessed 20 March 2015).

Hajer, M. A. (1995) *The politics of environmental discourse: ecological modernization and the policy process*. Oxford: Oxford University Press.

Interviews

Official at the Ministry of Agriculture, Ethiopia (2013) Semi-structured interview. Interviewed by Ethiopia research team from Echnoserve, 10 July 2013. (Transcript available upon request.)

Official at the Ministry of Power, Bangladesh (2013) Semi-structured interview. Interviewed by Nazria Islam, 4 November 2013. (Transcript available upon request.)

Official at the MOFE, Bangladesh (2013) Semi-structured interview. Interviewed by Nazria Islam, 24 October 2013. (Transcript available upon request.)

Klein, R. J. T., Huq, S., Denton, F., Downing, T. E., Richels, R. G., Robinson, J. B. and Toth, F. L. (2007) Inter-relationships between adaptation and mitigation. Climate change 2007: impacts, adaptation and vulnerability. In M. L. Parry, O. F. Canziani, J. P. Palutikof, P. J. van der Linden and C. E. Hanson (eds) *Contribution of Working Group II to the fourth assessment report of the Intergovernmental Panel on Climate Change*. Cambridge: Cambridge University Press, pp. 745–777.

OECD (2014) *Climate resilience in development planning: experiences from Columbia and Ethiopia*. Paris: OECD.

Nash, E. and Ngabitsinze, J. (2014) *Low-carbon resilient development in Rwanda*. IIED Country Report. London: IIED. Available at http://pubs.iied.org/10065IIED.html (accessed 12 March 2016).

Rai, N., Huq, S. and Huq, M. J. (2014) Climate resilient planning in Bangladesh: a review of progress and early experiences of moving from planning to implementation. *Development in Practice*, 24(4): 527–543.

4 National political economy of climate funds: case studies of the PPCR and the SREP

Neha Rai and Thomas Tanner

Contributing authors: Sunil Acharya, Ramesh Bhushal, Raju Chhetri, Md Shamshudoha, Misgana Elias Kallore, Sumanta Neupane and Lidya Tesfaye

Introduction

Over the last 20 years international climate finance has increased in scale, developed a complex architecture and drawn in a wide range of actors and institutions. Countries have made use of increasingly diverse sources of funding, financial instruments and intermediaries (Rai et al. 2015a; Kaur et al. 2014). Funding comes from both public and private sources, and may be channelled to developing countries in a variety of ways. These include funding mechanisms set up under the UNFCCC, as well as a range of multi- and bilateral channels operating outside the UNFCCC (Rai et al. 2015b). At the same time, recent developments in climate policy are driving the integration of the mitigation and adaptation agendas.

The increase in funds and diversification of funding mechanisms has led to changes in priorities and power relations at the national level. The picture is complex, given the interaction between the general international context, with its influence on climate finance incentives and governance, and the specific political environments of individual countries. While analysis of the international context has been plentiful (see, for example, Paterson and Grubb 1992; Luterbacher and Sprinz 2001; Aldy et al. 2003; Adger et al. 2006; Nakhooda and Norman 2014), there is a need for greater attention to be paid to this interaction, and to ask how international initiatives are being translated and reformulated in national contexts (Tanner and Allouche 2011; Naess et al. 2015).

In this chapter we address this question through case studies of the Pilot Program for Climate Resilience (PPCR) and the Scaling Up Renewable Energy Program (SREP) as they operate in Bangladesh, Ethiopia and Nepal. These two programmes represent global funding initiatives supporting the climate adaptation and climate mitigation agendas, respectively; both are part of the group of non-UNFCCC multilateral funds managed by the World Bank known as the Climate Investment Funds.

Using a political economy approach, our aim is to unpack the relationships between these international initiatives and national-level policymaking and processes concerning LCRD. In doing so, we are moving on from the early agenda-setting and formulation stages of the policy cycle discussed in Chapter 3 to consider decision making and implementation.

The analysis presented here is based on Hajer's (1995) discourse framework, as discussed in Chapter 1. This is concerned with how storylines or narratives supported by different groups of actors interact to generate consensus, cooperation, exclusion and competition in the policy process. These storylines are derived from the ideologies and incentives actors are subject to, based on their institutional roles, remits and structures, on the procedures and policies they follow, and on the resources and knowledge available to them. Storylines are used to support and justify a particular set of actions (Roe 1991; Leach et al. 2010). According to Hajer, storylines are 'devices through which actors are positioned, and through which specific ideas of "blame" and "responsibility" and "urgency" and "responsible behavior" are attributed' (1995: 64–65).

By exploring the influence of different storylines and decision-making coalitions in this way, our analysis reveals how national-level decisions about climate finance can be seen as the result of an ongoing renegotiation of ideas and ideology. It provides valuable insight into how international climate finance programmes are translated into national policies, shedding light on how and why governments adopt a particular course of action (or non-action) and how they put their policies into effect. By understanding institutional structures, powers and capacities, we can analyse the political and ideological processes underpinning climate finance and its governance.

Our findings reveal some common patterns in decision makers' engagement with the climate finance process and highlight some of the potential pitfalls. They lead us to make three key arguments:

- First, actors with shared storylines and resources form coalitions in support of certain investment decisions. Our case studies show how actors involved in planning investments under the PPCR and the SREP coalesced into groups based on particular shared visions of how to use these funds to achieve 'transformational change' and to realise particular development benefits within their countries. These discourse coalitions developed into more formal policy networks supporting investment decisions.
- Second, incentives that include policy, economic and knowledge-based factors can strengthen the discourse coalitions supporting particular decisions. We found that actors with sufficient resources and a pre-existing network of institutional relationships were well placed to put their storylines into practice, in the form of actions and investments that were consistent with their shared views.
- Third, more disparate storylines and storylines supported by actors lacking resources have less influence on decision making, but

nevertheless can undermine implementation. Our case studies provide examples of less mainstream views, voiced by stakeholders on the margins of debate, which failed to translate into investments. Even if they do not sway the policy consensus, however, such dissenters are often in a position to hinder projects they disagree with.

These findings lead us to argue that outcomes can be improved if national-level actors understand the internal political economy of the decision-making process and use this understanding to deliver plans with wide stakeholder support. Clarity about the political economy of climate investments can help leaders to harness consensus, avoid obstacles and choose more equitable, representative projects to pursue.

The international context: multilateral climate funds

In this section we examine two multilateral climate funding mechanisms used to deliver international climate finance, and we look at where the CIFs sit within this landscape. Multilateral climate funds can be broadly divided into those that operate under the aegis of the UNFCCC and those that operate outside it. UNFCCC funds include:

- the GCF – the UNFCCC's main financing mechanism
- the Global Environment Facility – originally set up as an interim measure
- the Adaptation Fund – founded under the Kyoto Protocol
- the UN initiative on Reducing Emissions from Deforestation and forest Degradation (UN-REDD).

Non-UNFCCC funds include:

- the CIFs, which include the PPCR and the SREP
- the Global Climate Change Alliance
- the Global Energy Efficiency and Renewable Energy Fund
- the Forest Carbon Partnership Facility.

Figure 4.1 illustrates the amounts delivered via these two funding mechanisms. The CIFs are clearly by far the largest non-UNFCCC source of climate funding.

Although both the UNFCCC and non-UNFCCC funds share the core aims of mitigating greenhouse gas emissions and building climate resilience, there are some key differences in the way in which they function. These differences and the incentives they create can have significant implications for countries' investment choices. For example, some UNFCCC funds allow accredited national agencies to access funds directly. This promotes a sense of ownership, with decisions being made at national level. In contrast,

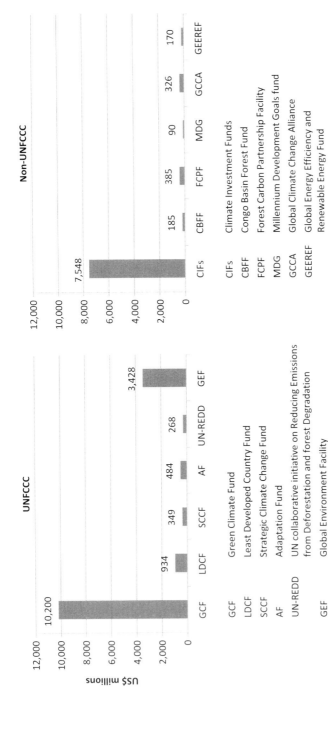

Figure 4.1 UNFCCC and non-UNFCCC multilateral climate funding
Source: Data obtained from the Climate Funds Update database (2015).

non-UNFCCC funds, including the CIFs, usually only grant access indirectly via multilateral agencies.

The financial instruments used also tend to differ. UNFCCC-operated funds have traditionally provided grants whereas non-UNFCCC funds, again including the CIFs, have deployed a wider range of instruments, including concessional loans, guarantees and private equity investments. In some countries, however, controversy has arisen around the use of specific instruments, with implications for the level of acceptance in these countries for the use of climate funds and funding systems. In the case of the CIFs, some developing countries and civil society organisations have expressed concern about the provision of loans for adaptation projects. The argument is that by providing loans rather than grants, developed countries effectively avoid paying the compensation they owe for excessive historical carbon emissions (Hulme et al. 2012).

UNFCCC funds such as the Adaptation Fund have largely taken a project-based and issue- or sector-driven approach to funding, which is made available on a first come, first served basis. This means that funding can be used to target smaller-scale community-based projects, and this has been effective in combination with these funds' provision of direct access and grant-based instruments. However, it also tends to encourage submission of existing project proposals by 'early mover' countries. This may have distributive implications, effectively excluding countries that are in need of funding but which have lower levels of 'readiness'. In contrast, the CIFs emphasise a programmatic approach that coordinates a range of projects around a common objective. This allows countries to plan large-scale efforts that link up with their national priorities.

The CIFs also provide readiness support to help countries to develop their proposals, as well as funding implementation. Funds are disbursed in two phases: a planning phase for the development of project proposals and institutional capacity building, followed by an implementation phase to put these plans into effect. The Adaptation Fund again presents a contrast, offering only project formulation funding to enable proposals for approved projects to be refined.

These differences in the nature of two funding mechanisms, along with the emergence of new players and institutions, are shaping national-level LCRD investments, sometimes in ways that may not be the most equitable, efficient or representative (Nakhooda and Norman 2014).

Case studies

The analysis presented here is based on empirical research carried out as part of an IIED study into the political economy of the PPCR and the SREP (see Rai et al. 2015b). Drawing on results from semi-structured interviews, we use discourse analysis to consider stakeholders' varying interpretations of the CIFs' core objectives, and the implications for how these objectives are operationalised. The overall aim is to gain insight into how international climate funds are negotiated and translated at national level, showing what happens when they hit the realities of politics and power relations 'on the ground'.

The core objectives set out for the CIFs include (CIFs 2009, 2010):

- effecting transformational change in recipient countries;
- contributing to development impacts in these countries;
- catalysing private sector involvement in climate change actions;
- ensuring countries have ownership of their climate change actions;
- scaling up investments.

The 'transformational change' of the first of these objectives can be characterised as a shift away from business as usual decisions at national level. It implies a long-term process of 'institutional and policy changes, technological shifts, and re-orienting investment priorities … to demonstrate effects, remove barriers and develop mechanisms for replication [of projects]' (ICF 2013: 18). The 'developmental impacts' of the second objective, meanwhile, refer to outcomes such as reducing poverty and improving health and education, particularly among the poor and more vulnerable.

The PPCR seeks to bring about the CIFs' core objective of 'transformational change' through a programmatic investment strategy (CIFs 2011; Rai et al. 2015b). Its overall aim is to create an integrated approach to climate change adaptation in low-income countries. The SREP, on the other hand, targets climate mitigation, providing finance to showcase investments in low carbon energy technologies in low-income countries (Rai et al. 2013).

Each case study begins with a general description of the climate finance programme concerned and the related administrative structures in the case study countries. We then move on to analyse the political economy of the investment decisions made in these countries. We examine the storylines, discourse coalitions, actors, and incentives – the policy, economic and knowledge-based factors – that have shaped these decisions. We also identify alternative and marginal storylines and assess their influence on decision making and project implementation.

Case study: the PPCR in Bangladesh and Nepal

The PPCR

The PPCR was established in 2009 with the aim of enabling low-income countries to develop an integrated, scaled-up approach to climate adaptation. It is the largest of any of the adaptation funds, with a total value (at 2015) of US $1.2 billion. Initially, 20 countries and two regions were chosen to receive funds under the programme. Funding is made available for capacity building and policy reform, long-term institutional strengthening through technical assistance, and investment in activities and assets on the ground. Investments are typically focused on one or two themes or subregions, and funding is usually delivered through a combination of grants and loans.

Like the other CIFs, the PPCR supports two phases of activity, planning and implementation. The planning phase involves a multi-sectoral dialogue to prioritise investments and create a national investment plan, termed a Strategic Program for Climate Resilience (SPCR). Typically, it also entails delineating the policies, strategies and development plans that need to be updated to achieve climate resilience, as well as dividing up key tasks between agencies – government bodies, development banks and other partners – and setting up a results framework to track progress (CIFs 2009). The implementation phase then involves operationalising the investment priorities set out in the SPCR.

As Figure 4.2 shows, the PPCR has a diverse project portfolio, with the three areas receiving the highest PPCR funding being agriculture and landscape management, infrastructure, and water resource management. The figure also illustrates the co-finance generated by PPCR funding, and it is notable that this varies considerably by sector. Particularly high levels of additional finance are available for coastal zone management and infrastructure, and comparatively little in the areas of climate information systems and 'enabling environment' – that is, building capacity for investment in climate adaptation projects (CIFs 2014).

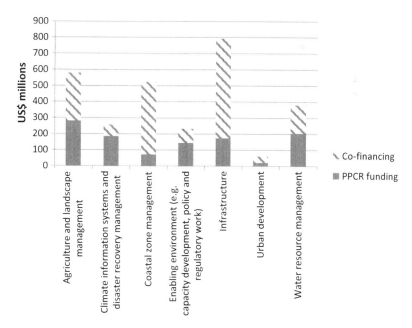

Figure 4.2 Total PPCR funding and co-financing in 2014, by investment type
Source: Rai et al. 2015a, available at http://pubs.iied.org/pdfs/10111IIED.pdf, © IIED.

Governance of PPCR funding varies from country to country, with some using existing administrative arrangements and others making new ones. New arrangements are more evident under the PPCR than the SREP, due to its dedicated technical assistance funding, used for 'mainstreaming' and upgrading administration and capacity development.

Some governments have shifted or shared core responsibility for climate change leadership beyond their environment ministries, typically to departments dealing with finance and planning. In Bangladesh, responsibility for PPCR activities is shared between the environment and finance ministries. The effect of this is to locate leadership with 'convening authorities' operating across multiple sectors. In a further demonstration of its wider buy-in, Bangladesh also co-finances PPCR projects and includes them in its annual development planning budgets. In contrast, Nepal has selected the MOSTE as its lead PPCR agency, despite external preference for the MOFED (Ayers et al. 2011; Rai 2013). Programmes prioritised in Nepal's SPCR (the PPCR proposal document mentioned previously) are then implemented by line ministries or other departments, in collaboration with multilateral development banks.

Bangladesh makes use of existing channels for implementation, in the form of government line departments and agencies already channelling considerable resources (Huq 2012; ICR 2013; Rai et al. 2014). The Bangladesh Water Development Board and the Local Government Engineering Department, for example, have received 45 per cent of their funding so far from the BCCRF and the BCCTF. These agencies are now responsible for implementing the two PPCR investment projects prioritised under the country's SPCR. They are also making use of long-standing partnerships with multilateral development banks. In Nepal, meanwhile, the two key implementation agencies are the Department of Soil Conservation and Watershed Management, and the Department of Hydrology and Meteorology.

PPCR investment decisions in Bangladesh and Nepal

In terms of their PPCR investment decisions and priorities, Bangladesh and Nepal have taken very distinct approaches (see Table 4.1). Bangladesh has used PPCR money to top up funding for the climate-proofing of its infrastructure, the aim being to increase the scope and scale of its adaptation actions. There has been limited focus on building institutional capacity through technical assistance. Accordingly, two substantial projects, pursued with funding from the Asian Development Bank and the World Bank, have prioritised coastal infrastructure. Key incentives here were the availability of co-financing from the banks (the amounts required were enormous and the available PPCR funds relatively small), and the pre-existing relationships and experience of the banks and government line departments in coastal infrastructure development. Thus investment decisions in Bangladesh were influenced by the government's ability to use concessional loans, raise co-finance and harness existing partnerships.

Table 4.1 PPCR investments in Nepal and Bangladesh

Country	Investment	Implementing agencies	
		Multilateral development bank	Government line department(s)
Bangladesh	Coastal Embankments Improvement Project (CEIP) and related afforestation project	World Bank	Bangladesh Water Development Board Forestry Department Bangladesh Forestry Research Institute
	Coastal climate resilient water supply, sanitation, and infrastructure improvement	Asian Development Bank	Local Government Engineering Department Department of Public Health and Engineering Ministry of Food and Disaster Management Water Supply and Sewerage Authority
	Promoting climate resilient agriculture and food security	International Finance Corporation	Department of Agricultural Extension within Ministry of Agriculture (initially) Ministry of Environment and Forests (later)
Nepal	Building climate resilience of watersheds in mountain ecoregions to improve agricultural productivity	Asian Development Bank	Department of Soil Conservation and Water Management
	Building climate resilience by supporting early warning systems for improved farming practices	World Bank	Department of Hydrology and Meteorology
	Mainstreaming climate risk management in development	Asian Development Bank	Ministry of Science, Technology and Environment (MOSTE)
	Building climate resilient communities through private sector participation	International Finance Corporation	
	Enhancing climate resilience of endangered species	World Bank	Ministry of Forest and Soil Conservation

Source: Adapted from Rai et al. (2015b).

In contrast to Bangladesh's infrastructure-heavy approach, Nepal has specifically targeted 'softer' capacity-building projects, including climate information systems for use in agriculture. The overall aim has been to improve communities' capacity to adapt to climate change by developing a variety of tools, instruments and strategies.

The main players in determining investment priorities in both countries were core government ministries such as the finance and environment ministries, line departments and multilateral development banks. Civil society organisations, development partners, agriculture ministry and multilaterals such as the UNDP were involved in Bangladesh's initial consultation process on the PPCR, but had limited input into prioritisation and programme delivery.

Discourse coalitions and narratives

Which narratives drove these distinct choices, and how were they shaped by the internal politics and incentive structures within these two countries? We analysed the views of stakeholders in both countries on how PPCR investments could contribute towards the CIFs' core objectives, specifically those of effecting transformational change, contributing to development impacts and catalysing private sector involvement. Based on this analysis we distinguished which were the 'dominant' narratives – supported by clear discourse coalitions and actually translated into policies and actions – and which were 'alternative' narratives – views that remained on the margins, or that had only diffuse support among stakeholders. The results are summarised in Table 4.2.

Within Bangladesh two broad narratives defined actors' interpretations of the PPCR's ability to bring about *transformational change*. The dominant narrative reflected a belief in investing in *climate resilient infrastructure*, meaning that the pre-existing role of government line departments in climate-proofing the country's infrastructure should be further supported and enhanced. This narrative was widely supported by actors directly involved in designing and delivering PPCR investments, including those from the multilateral development banks, core government ministries and executing line departments.

The second, alternative narrative was one of *social innovation*. Its proponents expressed support for innovative social development, community-based adaptation, and improving livelihoods and basic services among the poor. The actors supporting it were in general not directly engaged in the delivery of PPCR investments, however, coming from multilateral organisations such as UNDP, civil society groups, non-executing line departments and some core ministries (for example the planning ministry). They shared few points of contact or resources, and with a consequent lack of coalition and incentives to action diluting their influence, the alternative narrative thus had little effect on decision making.

Actors' views on the potential of PPCR investments to contribute to development in Bangladesh also reflected two broad storylines. Stakeholders

Table 4.2 Narratives and incentives related to PPCR investment decisions in Bangladesh and Nepal

	Bangladesh	Nepal
Investment decision	• Longer-term infrastructure investments, e.g. coastal embankments, water sanitation	• Capacity-building projects • Climate information and early warning systems
Dominant narratives	• Transformational change can be achieved by providing climate resilient infrastructure • Development impacts can be achieved by targeting economic growth	• Transformational change can be achieved by meeting long-term sustainability goals and increasing climate adaptation capacity • Development impacts can be achieved by focusing on social development
Actors supporting dominant narratives	• Core government ministries, e.g. finance ministry • Affiliated line ministries and departments • Multilateral development banks	• Core government ministries, e.g. finance ministry • Affiliated line ministries and departments • Some multilateral development banks
Alternative narrative	• Transformational change and development impacts can be achieved through socioeconomic innovation and inclusive development	• Transformational change and development impacts can be achieved through infrastructure investments and economic growth
Actors supporting alternative narrative	• Other government bodies • Civil society • Other multilaterals, including UN agencies	• Some multilateral development banks
Incentives shaping dominant narratives	*Economic incentives* • Projects already in pipeline • Track record in using technology for infrastructure projects • Existing partnerships between multilateral development banks and government line departments *Policy incentives* Investment priorities in existing climate change policies: • BCCSAP • NAPA *Knowledge incentive* • Results of vulnerability and loss and damage assessments	*Economic incentive* • Availability of concessional loans *Policy incentives* Investment priorities in existing climate change policies: • NAPA • Climate Change Adaptation Framework for Agriculture *Knowledge incentive* Research results: • Water scarcity • Ineffective forecasting systems

from multilateral development banks, some core implementing ministries and bilateral agencies supported a dominant narrative based on using PPCR funds for investments that should help to set the country on a path to *economic growth*. Actors not directly involved in PPCR decision making supported an alternative *social development* narrative, calling for investments focused on equity and inclusivity.

Storylines in Nepal were essentially the reverse of those in Bangladesh, with the dominant transformation narrative centring on *long-term sustainability* and developing capacity in climate resilience planning. In particular, core actors such as government officials supported this sustainability narrative, viewing PPCR investments as models for the future. Executing departments, meanwhile, tended to support a *capacity development narrative*, seeing PPCR as an opportunity to transform institutions and build technical expertise. Similarly, the dominant development storyline was concerned with inclusive social development, while stakeholders calling for *infrastructure development and growth*, principally from multilateral development banks, remained on the fringe. In general, alternative storylines in Nepal were more diffuse than those in Bangladesh, often reflecting no clear consensus or coalition.

Shared narratives play a central role in both investment decisions and programme implementation. Nepal's PPCR investments primarily focus on climate information systems and other capacity-building initiatives that in the coming decades will give local government and farming communities a firmer basis for their adaptation decisions. Discourse analysis of our interview data suggests that this capacity-building approach reflects a dominant narrative with broad support among government officials based on the transformational potential of long-term sustainability and increased capacity for climate adaptation. The country's PPCR pilot projects are seen as the first steps in a long-range approach prioritising inclusive social development. Alternative storylines with support among multilaterals call for an approach based on infrastructure development, growth and employment. These views have remained on the fringe.

While alternative storylines do not generally translate into policies and actions, our analysis illustrated that they can hinder the implementation of projects. For example, in relation to the PPCR's objective of catalysing private sector investment in adaptation-related activities, in both Bangladesh and Nepal the original intention was that responsibility for implementing this policy in the key areas of agriculture and food security would rest with the agriculture ministries. However, officials in these ministries proved reluctant to spend public funds to incentivise profit-oriented businesses. They argued that the private sector has insufficient capacity, its capabilities not extending much beyond the supply of basic goods such as seeds, fertilisers and pesticides. In both countries this led to disagreement between the implementing multilateral development bank – the International Finance Corporation (IFC) – and these key line ministries, causing significant delays in implementation. In both cases the solution devised by the IFC was to bypass the agriculture ministry. In Bangladesh the policy is now operated primarily by the environment ministry while in Nepal the IFC manages it directly.

Other examples where differences of opinion undermined implementation included Bangladesh's project to repair coastal embankments. This included plans to improve surrounding areas of forest; however, these have been scaled back due to a lack of cooperation between the country's Water Development Board (responsible for the coastal embankments) and its Forestry Department. Similarly, imbalances in the allocation of resources for Nepal's PPCR investments also appear to be creating dissent. The project to provide climate early warning systems, managed principally through the country's hydro-meteorology department, has been allocated a greater share of the funding. As a result, projects to develop climate resilient technologies and link farmers to an early warning system will receive much less funding than the agriculture ministry had originally envisaged (Rai et al. 2015b).

As these instances show, conflicting or marginalised views – views not integrated into policy or otherwise resolved – have the potential to set up 'roadblocks' to implementation. Where there is insufficient policy consensus, those with dissenting opinions can be in a position to disrupt plans they disagree with.

Incentives

Why did some narratives become dominant, while others did not feature in the decision-making process? Our evidence illustrates that a wide range of incentives supported the different discourse coalitions' choice of storylines, variously generating consensus, cooperation or exclusion. Table 4.2 shows the main incentives supporting the dominant narratives in the two case study countries.

Briefly, incentives can be categorised as policy, resource or knowledge incentives. Policy incentives include guidelines, regulations and institutional mandates. Economic incentives are provided by the availability of money, technologies, expertise and other resources. Knowledge incentives are provided by evidence and understanding, which then provide a basis for decision making.

Our analysis suggests that policy incentives played a crucial role in supporting dominant storylines in Nepal. These incentives were provided principally by two high-level strategic plans – the country's NAPA and a sectoral framework for adaptation in agriculture – which focus on the country's needs with respect to agriculture, water and climate information. Research results highlighting water scarcity and the poor quality of existing weather forecasting also represented knowledge incentives for investing in climate information systems.

In Bangladesh, economic and knowledge incentives were important factors in the decision to prioritise infrastructure investments. Climate change vulnerability assessments, including evaluations of loss and damage resulting from Cyclone Sidr in 2007, called for US $1.2 billion to rehabilitate coastal embankments. Economic incentives included the experience of multilateral development banks and line departments of working together on infrastructure projects; this encouraged the government to use PPCR money for similar purposes. In addition, co-finance was available for coastal infrastructure projects that were already in the banks' pipelines and ready for funding top-ups; PPCR represented a means of scaling up the level of investment.

National adaptation policies also supported Bangladesh's PPCR investment decisions. Its climate change strategic plan includes infrastructure improvement as one of six priorities for dealing with climate change impacts (Government of Bangladesh 2009). Similarly, its NAPA includes as a core priority 'enhancing resilience of urban infrastructure and industries to impacts of climate change' (Government of Bangladesh 2005).

As this analysis shows, the influence of these economic and non-economic incentives had the effect of strengthening shared narratives and policy networks, thereby shaping the investment decisions made.

Case study: the SREP in Ethiopia and Nepal

The SREP

Like the PPCR, the SREP was established in 2009. While the PPCR funds adaptation actions, the SREP focuses primarily on piloting renewable energy technologies in low-income countries, with the aim of demonstrating their economic, social and environmental viability (Rai et al. 2013). Its main target is investment, although it supports related policy reforms and capacity development. It also emphasises the role of the private sector in achieving a sustainable increase in the use of renewables.

The SREP provides funding in the form of grants, concessional loans, guarantees and equity. The planning phase of funding supports countries in developing an investment plan. Grants are provided for capacity building and advisory services, while loans support the cost of introducing new technologies. This is followed by the implementation phase (CIFs 2010).

Figure 4.3 shows SREP funding by country and type. The technologies which have received investment fall into two main categories: those aimed at providing energy access, in the form of mini-grids (which combine hydro, solar photovoltaics (PV) and wind energy), off-grid distributed solar PV technology and improved cooking stoves, and renewable grid-tied technologies, including geothermal, wind, solar PV, hydro and solar-wind hybrid, and waste-to-energy projects. At 2013 about 25 per cent of SREP funds had been used for energy access projects and 65 per cent for renewable grid-tied projects; the remaining 10 per cent was allocated to capacity building.

In most participating countries the energy ministry is the government department with overall responsibility for SREP involvement. However, in Nepal the finance and environment ministries share this role, while in Ethiopia it is part of the remit of the Environmental Ministerial Council.

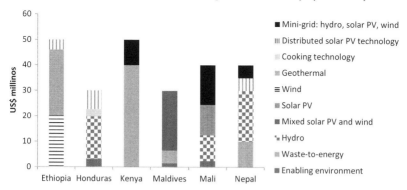

Figure 4.3 SREP total funding by country and technology type at 2013
Source: Rai et al. (2015a), available at http://pubs.iied.org/pdfs/10111IIED.pdf, © IIED.

This is an interministerial group that controls the country's CRGE Facility, of which SREP funding forms part.

The institutional architecture for implementing SREP projects often combines existing and new arrangements. In Nepal, it is the responsibility of the Alternative Energy Promotion Centre (AEPC), a semi-autonomous agency focusing on investment in small-scale renewables (AEPC 2013). A Central Renewable Energy Fund (CREF) and steering board have also been set up to mobilise funds from different sources – government and donors, and grants and loans. Its remit also includes engaging with the private sector. A dedicated agency administers decentralised energy projects of up to 10 megawatts, while larger-scale energy projects are the responsibility of the energy ministry.

In Ethiopia, the Ministry of Water, Irrigation and Energy (MOWIE) and the CRGE Facility are the key implementing bodies, with MOWIE taking the lead through its SREP coordination unit. Geothermal and wind projects are implemented by the Ethiopian Electric Power Corporation and the Ministry of Mines, in collaboration with a multilateral development bank.

In both countries, the IFC aims to catalyse private sector investment in renewable energy by providing incentives to commercial banks. In Nepal, the private sector has been involved in choosing SREP investment projects, under the management of the CREF and a commercial bank. The IFC and the Asian Development Bank aim to incentivise the private sector to invest in grid-connected renewables. As with the PPCR, the private sector component of the SREP involves direct agreements with multilateral development banks, and limited engagement from public bodies.

SREP investment decisions in Ethiopia and Nepal

Table 4.3 shows SREP investments in Nepal and Ethiopia; again, their respective approaches are quite distinct. Ethiopia's overall aim is to fuel economic growth by scaling up and diversifying the energy supply – currently the country

Table 4.3 SREP investment portfolio, Ethiopia and Nepal

Country	Investment	Implementing agencies	
		Multilateral development bank(s)	Government agencies
Ethiopia	Aluto Langano Geothermal Field Development and Geothermal Sector Strategy	African Development Bank	Ministry of Mines Environmental Protection Authority initially, now Ministry of Environment and Forest following restructuring
	Assela Wind Farm Project	World Bank	Ethiopian Electric Power Corporation
	Clean Energy SMEs Capacity Building and Investment Facility	International Finance Corporation	Ministry of Water, Irrigation and Energy (MOWIE) Ministry of Finance and Economic Development (MOFED)
Nepal	Small on-grid solar project	Asian Development Bank International Finance Corporation	Department of Soil Conservation and Water Management
	Mini and micro initiatives: off-grid solar PV and mini or micro hydropower	Asian Development Bank	AEPC
	Extended Biogas: scaling up waste-to-energy generation	World Bank	AEPC

Source: Adapted from Rai et al. (2015b).

is heavily reliant on hydropower, which is particularly sensitive to climate change. Accordingly, a large proportion of investment has gone into large-scale on-grid geothermal and wind energy. Three main projects have received funding: the Aluto Langano Geothermal Field Development and Geothermal Sector Strategy, which aims to help Ethiopia to achieve its target of generating 1 gigawatt of energy from geothermal sources by 2030; the 100-megawatt Asseia Wind Farm Project; and the Clean Energy SMEs Capacity Building and Investment Facility. The aim of the last is to 'skill up' and support women to run clean energy small and medium-sized enterprises (SMEs), thus removing barriers to the adoption of products such as cooking stoves for the home, mini-grids and Solar Home Systems (Rai et al. 2015b). However, this SME component is fairly small compared to the other two projects.

By contrast, SREP investments in Nepal are aimed at expanding energy access in remote areas using a variety of technologies, including hydropower, solar, wind and waste-to-energy. Its mini and micro initiatives are aimed at providing affordable energy access to rural populations, in the form of mini and micro hydropower installations and Solar Home Systems, with a total of 30 megawatts of generation capacity. An Extended Biogas project focuses on scaling up municipal waste-to-energy generation through covering initial costs and removing credit barriers. A small on-grid component is also included; originally planned as a hydropower project it has now been revised to use solar power.

Discourse coalitions and narratives

With regard to the PPCR, we analysed stakeholder interviews to identify the narratives and discourse coalitions supporting SREP investment choices. Stakeholders were asked for their views on how SREP investments could contribute towards the CIFs' objectives of achieving transformational change and development impacts. Table 4.4 provides a summary of our findings.

In Ethiopia, actors' views on the objective of transformational change and development coalesced around a dominant narrative of *economic growth*, driven by improving energy security through *diversification of on-grid technologies*. This view was supported by core and implementing government officials and those from multilateral development banks. Broadly, it reflected a belief that shifting away from business as usual hydropower to use a greater mix of energy sources was a transformative move that would reduce reliance on a single, vulnerable energy source and drive economic growth. An alternative narrative was articulated by marginal actors from bilateral funders, who argued that *improving energy access* for the rural poor would bring co-benefits for both people and the environment, for example by helping to prevent deforestation, to improve health and to reduce indoor pollution.

Priorities in Nepal were very different, and views were also more diffuse. Overall, government policymakers and multilateral development banks

Table 4.4 Narratives and incentives related to SREP investment decisions in Ethiopia and Nepal

	Ethiopia	Nepal
Investment decision	Large-scale on-grid geothermal, wind and solar energy	Rural small-scale hydropower (on- and off-grid), solar and wind energy, and waste-to-energy generation
Dominant narrative	• Transformational change can be achieved through diversification of energy technologies and economic growth. • Development impacts can be achieved through economic growth and improved employment opportunities.	• Transformational change can be achieved through transitioning to low carbon growth, with co-benefits for health, education and employment.
Actors supporting dominant narrative	Implementing line ministries Multilateral development banks Bilateral donors	Core government ministries Multilateral development banks
Alternative narrative	• Development impacts based on poverty reduction can be achieved by investing in energy access in rural areas.	• Transformational change and development impacts can be achieved by promoting innovative technologies and by scaling up supply.
Actors supporting alternative narrative	Bilateral donors	Varied set of actors
Incentives shaping dominant narrative	*Economic incentives* • Availability of co-finance • Income from energy export • Economic benefits of scaling up energy supply *Policy incentives* • Growth and Transformation Plan • National target of producing 1 gigawatt (6%) of electricity from geothermal sources by 2020 • National target of achieving status of a middle income country by 2025 *Knowledge incentive* • Knowledge of the negative impact of climate variability on hydropower, currently Ethiopia's main energy source	*Economic incentives* • Proven technologies and existing systems in place • Energy for productive uses • Commercially viable technology *Policy incentive* • National Rural Renewable Energy Programme focus on energy access in rural areas *Knowledge incentives* • Knowledge and long-term experience of specific technologies. • Multilateral development banks' expertise, based on experiences elsewhere

adhered to a narrative of transformational change and development through improved *energy access and poverty relief*. Investment decisions based on this view were focused on up-scaling proven technologies to improve household energy access, and on realising associated co-benefits in areas such as health, education and employment. A narrative of *economic growth and technology diversification* – similar to the dominant view in Ethiopia – was supported here by a limited number of stakeholders from bilateral funders and multilateral development banks, who were particularly interested in initiating a new waste-to-energy industry in Nepal.

These competing narratives translated into a mix-and-match of investments in Nepal, involving on- and off-grid technologies of different types. In the absence of a dominant consensus and of strong policy networks there were disagreements between actors and delays in implementation. In particular, an initial plan for multilateral investment in small-scale hydropower (both on- and off-grid) was replaced with grid-tied solar projects. Similarly, pilot projects involving new systems such as hybrid solar-wind and extended biogas have been controversial, partly because the technology is unproven but also because they are funded through concessional loans rather than grants. Nepal has historically objected to the use of loans to fund climate-relevant projects, holding the view that climate actions should be paid for by developed countries, as the original polluters.

Despite the SREP's core aim of catalysing private sector investment, 90 per cent of Ethiopia's SREP funds are being channelled to public sector projects. Views on private sector engagement have been divided in Ethiopia, with some private sector stakeholders adhering to narratives that would set private companies at the centre of the country's transformation plans, arguing that large-scale electrification projects are beyond the capabilities of public agencies. However, the view that has prevailed reflects the government's scepticism about the readiness of private companies to take the lead, and its related preference for nurturing the private sector on a smaller, more local scale.

As a result, the private sector component of the SREP in Ethiopia is covered by the remaining 10 per cent of funding, and has been allocated for building capacity among commercial banks and SMEs. It has also faced delays and lost momentum due to regulatory barriers to the involvement of commercial banks (Rai et al. 2015b). In contrast, Nepal's SREP investments are split 50-50 between public and private sector projects. This is largely because the IFC (which manages the private sector component of the CIFs) has opted for commercially viable grid-based investments in Nepal, also because there are fewer regulatory constraints on the private sector there.

Incentives

Turning to consider the incentives underpinning the investment decisions made, our evidence suggests economic and policy incentives were crucial in influencing Ethiopia's large-scale grid-based approach to SREP investments (see Table 4.4). Economic incentives included the availability of co-finance for geothermal projects and the policy objective, set out in the country's growth and transformation plan, of a fast-growing national grid that enables energy export, and so contributes to economic growth and the target of Ethiopia becoming a middle-income country by 2025.

The associated decision to diversify the energy technologies used has been shaped by knowledge incentives. Policymakers in Ethiopia are aware that existing energy sources – principally hydropower – are vulnerable to (indeed are already being affected by) climate variability. Also operating here is the policy incentive represented by the country's target of producing 1 gigawatt of power from geothermal sources by 2020.

In Nepal, the economic incentive of lower costs is behind government decisions to opt for proven and commercially viable technologies, while the choice of smaller-scale systems reflects the country's policy focus on providing power for the rural economy. More recently, funding partners have sought to encourage the private sector to move into novel technologies, particularly solar energy. Supplementing this are projects co-financed by multilaterals that invest in piloting waste-to-energy and hybrid solar-wind technologies. The incentives operating here are also economic, taking the form of the finance made available by donors.

Conclusions

The political economy analysis we have presented in this chapter illustrates how different narratives – supported by groups of actors influenced by different combinations of incentives – interact to generate consensus, cooperation, exclusion and competition in the policy process. This provides valuable insight into how international climate finance programmes are translated into national policies and actions.

We can see that investment decisions are the result of discourse coalitions and policy networks forming around shared ideas and resources. In Ethiopia, for instance, SREP investments prioritised grid-based renewables owing to the view held among powerful stakeholders that diversification of energy technologies would help to promote economic growth. In Bangladesh, PPCR funding decisions were driven by a widespread belief that building climate resilient infrastructure would provide a pathway to transformational change and development.

These coalitions are shaped by a range of incentives, including economic factors, policy goals and factual evidence. Ethiopia's SREP investment decisions reflect the country's targets for energy generation and the economic incentive of the co-finance available for renewable energy projects. The consensus about PPCR funding decisions in Bangladesh was boosted by factors such as the country's existing expertise in infrastructure projects and evidence provided by loss and damage assessments of the need for investment.

Our analysis has also highlighted the existence of alternative and non-mainstream views. In Bangladesh a range of actors expressed the belief that investment of PPCR funds in community-based adaptation measures aimed at social development would yield better results than business-as-usual infrastructure investments. In the absence of strong coalitions with clear incentives these actors were largely excluded from the decision-making process, with the result that their views were not reflected in investment decisions.

This does not mean, however, that alternative views have no influence. As illustrated by the limited success in recruiting the private sector into SREP programmes in Ethiopia and Nepal, stakeholders with competing views can significantly delay and disrupt action. It is also not necessarily the case that one narrative will always win out in practice; it is possible for differing views to be reflected in investment decisions. The use of SREP funding for a mixture of proven and new technologies in Nepal – for example mini and micro hydropower and solar systems as well as waste-to-energy projects – is one instance of this.

Based on these findings, we suggest that in order to reach effective decisions about climate funding and avoid barriers to implementation, governments and international climate finance initiatives need a thorough understanding of the internal political economy of these decisions. By mapping the interactions among various narratives and their supporting stakeholders, policymakers will be better able to manage expectations and risks, prioritise more equitable projects and fashion a workable consensus. Any given programme or project proposal will have its proponents and opponents, with views shaped by the particular incentives in play. To steer towards a broadly supported consensus and avoid time-consuming disputes, governments and development partners will need to be bold and find pathways that work in the context at hand.

A proactive approach is needed here: policies are likely to be more effective if decision makers have actively sought out and integrated diverse views. Often this is likely to mean reshaping incentives, for example by providing resources to support actors with dispersed but useful alternative views, or by supporting consensus-based coalitions. More thought needs to be given to the sequencing of decisions and the representation of actors in the policy process, for instance to ensure that actors previously excluded from decisions that directly affect them are actively involved, thus increasing their sense of ownership as well as promoting cooperation.

Policymakers rolling out new funding programmes need to recognise patterns of agreement and dissent. If actors share a vision, channelling resources in that direction can harness synergies. When actors hold different views, policy- and resource-based incentives can be used to integrate these views and achieve a level of consensus. If competing or dissenting views seem likely to create obstacles to implementation, it is important to negotiate and manage expectations.

References

Adger, W. N., Pavoola, Y., Huq, S. and Mace, M. J. (eds) (2006) *Fairness in adaptation to climate change*. Cambridge, MA: MIT Press.

AEPC (2013) *Central Renewable Energy Fund (CREF): financial intermediation mechanism* (Final Draft). Kathmandu: AEPC.

Aldy, J., Ashton, J., Baron, R., Bodansky, D., Charonvitz, S., Dirringer, E., Heller, T., Pershing, J., Shukla, P. R., Tubiana, L., Tudela, F. and Wang, X. (2003) *Beyond Kyoto: advancing the international effort against climate change*. Arlington, VA: Pew Centre on Global Climate Change. Available at http://www.c2es.org/publications/beyond-kyoto-advancing-international-effort-against-climate-change (accessed 26 March 2016).

Ayers, J., Anderson, S. and Kaur, N. (2011) Negotiating climate resilience in Nepal. *IDS Bulletin*, 42: 70–79.

Climate Funds Update (2015) *Climate Funds Update database*. Available at http://www.climatefundsupdate.org/data (accessed 19 January 2016).

CIFs (2009) *Programming and financing modalities for the SCF targeted program, the Pilot Program for Climate Resilience (PPCR)*. Washington, DC: CIFs. Available at http://www.climateinvestmentfunds.org/cif/sites/climateinvestmentfunds.org/files/PPCR_Programming_and_Financing_Modalities.pdf (accessed 26 March 2016).

CIFs (2010) *SREP programming modalities and operational guidelines*. Climate Investment Funds. Available at http://www.climateinvestmentfunds.org/cif/node/2387 (accessed 26 March 2016).

CIFs (2011) *Climate Investment Funds: PPCR*. CIFs. Available at http://www.climateinvestmentfunds.org/cif/sites/climateinvestmentfunds.org/files/CIF-PPCR%20Sept%20'11%20Final.pdf (accessed 26 March 2016).

CIFs (2014) *PPCR semi-annual operational report*. CIFs. Available at https://www.climateinvestmentfunds.org/cif/sites/climateinvestmentfunds.org/files/PPCR_15_3_PPCR_semi_annual_operational_report_rev.1.pdf (accessed 9 December 2015)

Government of Bangladesh (2005) *National Adaptation Programme of Action (NAPA)*. Dhaka: MOEF. Available at http://unfccc.int/resource/docs/napa/ban01.pdf (accessed 19 March 2016).

Government of Bangladesh (2009) *Bangladesh Climate Change Strategy and Action Plan 2009*. Dhaka: MOEF. Available at http://cmsdata.iucn.org/downloads/bangladesh_climate_change_strategy_and_action_plan_2009.pdf (accessed 19 March 2016).

Hajer, M. A. (1995) *The politics of environmental discourse: ecological modernization and the policy process*. Oxford: Oxford University Press.

Hulme, M., O'Neill, S. and Dessai, S. (2012) Is weather event attribution necessary for adaptation funding? *Science*, 334: 764–765.

Huq, S. (2012) Mainstreaming climate change adaptation into national planning. *The Daily Star*, 21 October. Available at http://www.thedailystar.net/news-detail-254618 (accessed 26 March 2016).

ICF (2013) *Independent evaluation of the Climate Investment Funds*. Submitted to: Evaluation Oversight Committee for the Independent Evaluation of the Climate Investment Funds. Washington, DC: ICF. Available at http://www.cifevaluation.org/cif_interm_report.pdf (accessed 26 March 2016).

Kaur, N., Rwirahira, J., Fikreyesus, D., Rai, N. and Fisher, S. (2014) *Financing a transition to climate-resilient green economies*. IIED Briefing Paper. London: IIED. Available at http://pubs.iied.org/17228IIED.html (accessed 22 December 2015).

Leach, M., Scoones, I. and Stirling, A. (2010) *Dynamic sustainabilities: technology, environment, social justice*. London: Earthscan.

Luterbacher, U. and Sprinz, D. F. eds (2001) *International relations and global climate change*. Cambridge, MA: MIT Press.

Naess, L. O., Newell, P., Newsham, A., Quan, J. and Tanner, T. M. (2015) Climate policy meets national development contexts: insights from Kenya and Mozambique. *Global Environmental Change*, 35: 534–544.

Nakhooda, S. and Norman, M. (2014) *Climate finance: is it making a difference? A review of the effectiveness of multilateral climate funds*. London: ODI. Available at http://www.odi.org/sites/odi.org.uk/files/odi-assets/publications-opinion-files/9359.pdf (accessed 19 March 2016).

Paterson, M. and Grubb, M. (1992) The international politics of climate change. *International Affairs*, 68: 293–310.

Rai, N. (2013) *Climate Investment Funds: understanding the PPCR in Bangladesh and Nepal*. IIED Briefing Note. London: IIED. Available at http://pubs.iied.org/pdfs/17151IIED.pdf (accessed 19 March 2016).

Rai, N., Appunn, K., Kaur, N. and Smith, B. (2013) *Climate Investment Funds: Scaling up Renewable Energy Programme (SREP) in Ethiopia – a status review*. IIED Country Report. Available at http://pubs.iied.org/10053IIED.html (accessed 19 March 2016).

Rai, N., Huq, S. and Huq, M. (2014) Climate resilient planning in Bangladesh: a review of progress and early experiences of moving from planning to implementation. *Development in Practice*, 24: 527–543.

Rai, N., Kaur, N., Greene, S., Wang, B. and Steele, P. (2015a) *Topic guide: a guide to national governance of climate finance*. London: Evidence on Demand. Available at http://www.evidenceondemand.info/topic-guide-a-guide-to-national-governance-of-climate-finance (accessed 22 December 2015).

Rai, N., Acharya, S., Bhushal, R., Chettri, R., Shamshudoha, Md., Kallore, M. E., Kaur, N., Neupane, S. and Tesfaye, L. (2015b) *Political economy of international climate finance: navigating decisions in PPCR and SREP*. IIED Working Paper. Available at http://pubs.iied.org/pdfs/10111IIED.pdf (accessed 22 December 2015).

Roe, E. (1991) Development narratives, or making the best of blueprint development. *World Development*, 19(4): 287–300.

Tanner, T. and Allouche, J. (2011) Towards a new political economy of climate change and development. *IDS Bulletin*, 42: 1–14.

5 Designing climate finance systems in Ethiopia and Rwanda

Nanki Kaur, Daniel Fikereysus, John Rwirahira, Lidya Tesfaye and Simret Mamuye

Introduction

National and international policymakers promote investment in LCRD in order to address the challenges of climate change and development. The stimulation of this investment is reliant on policies that will mobilise and deliver appropriate finance, including scaled-up finance to support current and projected costs; long-term finance to sustain and incentivise investment; and flexible and inclusive finance that is accessible to the most vulnerable.

Framing and implementing policy options to finance LCRD investment is a complex task. Achieving adaptation and mitigation synergies requires programmatic rather than project-based, siloed funding. There is a need for an adequate and sustainable scale of finance for cross-sectoral LCRD approaches. This in turn requires innovative financing mechanisms, with tailored and appropriate instruments to create synergies between adaptation and mitigation agendas. Both public and private funding will be needed to bring about an integrated low carbon and climate resilient future (Boyle 2013).

Currently, however, there are a number of issues constraining the financing of LCRD priorities, particularly in low-income countries.

- First, the current supply of finance does not meet the LCRD needs of low-income countries. There is also an inequitable distribution of funds between low carbon and adaptation flows, with a significant proportion being spent on mitigation actions in emerging economies.
- Second, current practice tends to maintain a false dichotomy between mitigation and adaptation actions. One reason for this is that maintaining distinct funding streams makes it easier for development partners to demonstrate and attribute outcomes. As a result there are relatively few examples of how to provide finance for more integrated actions.
- Finally, current climate finance does not match up with the cost and time frames of LCRD investments and it is not easily accessible to the most vulnerable countries and communities. The LDCs require long-term finance for enabling activities, innovative technology and long-term investments, and for integrating the adaptation, mitigation and

development agendas. However, there is a lack of appropriate financial instruments and mechanisms for targeting these objectives.

These ongoing issues, along with climate-induced uncertainty, are creating a shifting landscape for policymakers. Negotiating this landscape successfully means making continuous adjustments, identifying and managing options using an iterative, learning-by-doing approach. Policy actors at national and international level are attempting to address the issues in various ways. For instance, government agencies in Ethiopia and Rwanda are exploring how to work with national financial institutions to mobilise and deliver finance to small-scale LCRD investors (Rai et al. 2015; Steinbach et al. 2015; Kaur et al. 2016). Actors are also using a range of financial management systems to govern the flow of finance; the same government agencies in Ethiopia and Rwanda plan to use public finance management systems to manage and allocate climate finance for investment in pro-poor public goods. Actors are also deploying a range of financial instruments, from grants to equity and debt finance, to incentivise investment in LCRD.

A number of lessons can be drawn from the experience of countries that have engaged with the LCRD agenda. In this chapter we use a political economy approach to examine policy choices related to climate finance, with a particular focus on the actors and policy networks involved in making these choices. We examine how two LDCs have engaged with specific financial intermediaries, instruments and planning systems to access and allocate climate finance from different sources. Our research included a review of literature, analysis of policy documents and semi-structured interviews with a range of government and donor stakeholders. We use the results here to analyse the incentives driving the formation of policy networks, in order to understand the factors influencing policy choices.

We have chosen to focus on Ethiopia and Rwanda because these two countries were among the first in Africa to adopt LCRD policies as their chosen route to achieving national development objectives (Fisher et al. 2014). Ethiopia's CRGE initiative is intended to transform development planning, investment and outcomes, and to provide a pathway for the country to achieve middle-income status by 2025 (Fikreyesus et al. 2014). Rwanda's NSCCLCD is intended to create a developed, low carbon and climate resilient economy by 2050. Both strategies are designed specifically to deliver inclusive and sustainable development.

Our analysis leads us to draw some core conclusions about the kinds of policy choices countries have made in order to meet their LCRD priorities. First, there are two main approaches used to mobilise and deliver LCRD finance: a more 'traditional' single-channel modality and a newer multi-channel modality, each supported by different coalitions. Second, factors that incentivise actors to follow a specific approach include path dependence, the existence of pre-existing relationships and experience of using particular instruments and processes; institutional mandates may also be relevant. Finally, the nature of the LCRD agenda incentivises policy actors to establish entities that use a multi-channel modality, with mandates to mobilise and blend finance from different sources in order to fund long-term, programmatic investments requiring cross-sectoral cooperation.

Our evidence suggests that an understanding of actors and their policy approaches, in synergy with an understanding of the policy networks and factors that support these approaches, can help to improve coordination between policy actors implementing LCRD finance. We begin our analysis by mapping out the LCRD investment landscape.

Understanding the LCRD investment landscape

In this section we describe LCRD investment priorities in Ethiopia and Rwanda, as set out in their respective national policies, and use their projected costings to consider the extent to which current climate finance flows match up with their investment needs. We show that in both countries there is a disjunction between national investment priorities and the scale and type of finance available.

LCRD investment priorities

In Ethiopia, low carbon, climate resilient investment priorities are guided by the country's GTPs, CRGE Vision, Green Economy Strategy and Climate Resilient Strategy. The CRGE Vision is for a climate resilient green economy by the year 2025. The Green Economy Strategy takes an economy-wide approach to achieving development goals while reducing greenhouse gas emissions, with the specific target of limiting them to current levels by 2030 (FDRE 2011). The Climate Resilient Strategy takes a sectoral approach to prioritising investment in climate-sensitive sectors. The government is currently developing guidelines that will integrate climate actions into Ethiopia's national development planning process (as set out in its GTP II). Guidelines for developing CRGE investment plans categorise investments into three types:

- Type 1 – enabling activities: interventions aimed at enabling investment in CRGE, such as climate information services and climate legislation
- Type 2 – mainstreaming activities: interventions aimed at mainstreaming CRGE into regular development and economic investments
- Type 3 – investments: these are made in addition to mainstream development investments; an example might be investments in renewable energy technology.

Investment priorities are also evident in Ethiopia's Fast Track Initiative. This is designed to kick-start the implementation of CRGE initiatives in prioritised sectors, including agriculture, forestry, energy, industry, transport and urban development. Each investment is expected to contribute to the triple objectives of encouraging economic growth, reducing greenhouse gas emissions and building resilience to climate change (Kaur et al. 2016).

In Rwanda, investment in LCRD is guided by the country's Economic Development and Poverty Reduction Strategies (EDPRS 1 and 2) (RSB 2013) and the previously mentioned NSCCLD (REMA 2011). The EDPRS provides policy direction for a green approach to economic transformation. It incentivises investment in sustainable cities and villages and in green innovation in industry and the private sector. The NSCCLD prioritises investment in low carbon energy supply and land use and water resource management, as well as social protection, improvements to health and disaster risk reduction.

Rwanda has also set up a national environment fund called FONERWA, the activities of which indicate the country's investment priorities. Funds are disbursed to investments falling within four FONERWA 'financing windows' (REMA 2010): conservation and management of natural resources; research and development, and transfer and implementation of technology; environment and climate change mainstreaming; and environmental impact assessment, monitoring and evaluation (Kaur et al. 2016).

FONERWA uses a multi-channel modality, mobilising resources from multiple agencies which are then channelled through a range of public and private sector entities, including civil society organisations and academic institutions. Public sector recipients of FONERWA climate finance include government agencies and districts, with 10 per cent of total funds earmarked for districts. The private sector, meanwhile, receives 20 per cent.

In both countries, investment in LCRD is expected to be made by both the public and the private sector. For instance, in Rwanda district government and sector ministries are expected to budget for LCRD investment, while investment in LCRD by private enterprises is projected to increase over time.

Financial needs and gaps

Costings produced by both countries provide estimates of the funding required for their respective LCRD investment plans. In Ethiopia, green economy investments are expected to require US $150 billion by 2025, or approximately $7.5 billion per year. Climate change adaptation is forecast to cost up to $10 billion per year. The country receives finance from domestic sources and overseas development assistance; however, demand exceeds the current supply. In the budget year 2011/12 an estimated $569 million (1.8 per cent of GDP) was allocated to cover climate-relevant expenditure (Eshetu et al. 2014). Taking a more 'macro' perspective, domestic revenue in the five-year plan for 2010–15 is expected to show an increase from $3.5 billion in 2010 to $9.15 billion in 2015 (MOFED 2010). International sources of finance provided a total of $1.3 billion over the four years from 2010 to 2013, in the form of overseas development assistance for investment in adaptation and mitigation (OECD-DAC 2013). Based on these figures, it is clear that the country will not be able to find the $7.5 billion per year required for CRGE initiatives (Kaur et al. 2016).

In Rwanda, the financing gap (across all the FONERWA thematic windows) is estimated at US $100 million per year. This assessment is based on differences between requested and approved budgets in the country's 2010/11 budget law, as well as financing gaps identified in sector and subsector strategic plans. Rwanda has mobilised $3.7 million to date from government sources and $357.69 million over four years (2010–13) from ODA (overseas development assistance) for investments in adaptation and mitigation (OECD-DAC 2013).

Ethiopia and Rwanda aim to support investment in LCRD for the next ten to 35 years. This will require access to sustained sources of finance. Also, the types of investment being prioritised, which include enabling activities, new and renewable sources of energy and innovative technology, will need to draw down on sources of finance that have a long-term investment time frame, such as sovereign wealth funds and institutional investors. However, both countries have so far drawn down on national and international sources of public finance that are subject to political timescales, with only short- to medium-term investment horizons; most ODA commitments, for example, have a one- to five-year horizon. We can see, then, that the current supply of finance is inadequate in terms of both scale and type.

This mirrors a mismatch between demand and supply in global flows of climate finance. Currently, these are dominated by short-term sources of private finance, which are mostly invested in the country of origin (that is, in developed countries) with the aim of securing an economic return on investment (see Table 5.1). As such, many of the LDCs, and poor and small-scale investors within these countries, lack access to scaled-up and long-term finance for LCRD.

Shaping the national climate finance landscape: actors, policy options and policy networks

Given the investment landscape described, policymakers need to identify appropriate financial intermediaries, instruments and planning systems to manage existing public finance in ways that will leverage and channel scaled-up, long-term finance for inclusive investment in LCRD. In this section we explore how Ethiopia and Rwanda are designing policy options to address this issue, and how effective these choices are. We then go on to consider the clusters of actors supporting these choices.

Policy actors influencing climate finance in Ethiopia and Rwanda

A range of international and national actors are influencing policy choices in the national climate finance landscape in Ethiopia and Rwanda. International actors include multilateral banks and agencies such as the World Bank, the

Table 5.1 Global flows of climate finance

Source	Intermediary	Instrument	Use
Global climate finance flows reached approximately US $343–$385 billion in 2010/11.	Public and private intermediaries, especially national development banks and commercial banks, played an important role in raising and channelling global finance of US $110–$120 billion.	Most climate finance – US $293–$347 billion out of the $364 billion – can be classed as investments in which public or private financial institutions have ownership interest or claim, i.e. money which has to be paid back, rather than contributions to incremental costs. Major categories of instruments include policy incentives (including income-enhancing mechanisms such as feed-in tariffs, tradable certificates, tax incentives and subsidies); risk management (including guarantees); carbon offset finance (under the Clean Development Mechanism); grants; low-cost debt; and capital instruments (including project-level market rate debt and equity and balance sheet financing).	Mitigation activities (mostly in the realm of renewable energy and energy efficiency) received US $350 billion and adaptation activities received $12.3–$15.8 billion (primarily as incremental cost payments). Agriculture and forestry received the greatest share.
The private sector contributed US $217–$243 billion or 63% of the total amount (the majority of the finance), which mostly came from asset finance.	Private commercial banks and infrastructure funds intermediated approximately US $38 billion.	Market-rate loans and equity provided US $293 billion.	
The public sector contributed US $16–$23 billion or 5–6% of the total amount. A large proportion came from national governments and ODA.	A large part of the ODA was received via bilateral finance institutions. Dedicated climate funds contributed at least US $1.5 billion to overall flows.	Concessional loans provided 60% of public finance, grants provided 7%. Risk management instruments (guarantees), grants, low-cost debt and balance sheet financing are the most common forms of finance provided by government money.	A sizeable proportion of domestic public finance was used to support renewable energy and related infrastructure.

Source: Data extracted from Buchner et al. (2013).

African Development Bank and UNDP. Policymakers from these institutions are responsible for managing multilateral sources of climate finance disbursed to Ethiopia and Rwanda. For instance, the African Development Bank and the IFC are responsible for managing SREP funds allocated to Ethiopia, while UNDP is responsible for managing the international window of Ethiopia's CRGE Facility.

International climate funds such as the GCF, the Adaptation Fund and the CIFs represent another group of international actors. Unlike individual multilateral entities, these funds pool resources from a wide range of multilateral and bilateral sources and channel them through multiple entities, including multilateral development banks, government agencies, private sector actors and civil society organisations. They can also use a range of instruments, such as concessional loans, grants and risk guarantees.

Bilateral partners such as the UK's DFID, Norway's Norad, Denmark's Danida and the Austrian Development Agency have supported the operationalisation of Ethiopia's national climate change fund. Representatives of all three organizations serve on the advisory board of the CRGE Facility, and DFID and Norad are supporting the use of public finance management systems and results-based financing. Similarly, in Rwanda DFID, Germany's KfW Development Bank and the Dutch Ministry of Foreign Affairs have contributed to the operationalisation of FONERWA.

Other international non-governmental and intergovernmental organisations, such as the Global Green Growth Institute and the Climate and Development Knowledge Network, provide support to federal ministries, the CRGE Facility and FONERWA on issues such as capacity building, knowledge management and the preparation of strategy documents.

At national level, development finance institutions, national climate change funds and commercial financial institutions are involved in LCRD financing. National agencies such as the finance ministry and key line ministries responsible for disbursing climate finance are responsible for developing fiscal policy and the public finance management systems used to mobilise and manage funds.

National development finance institutions act to mobilise and deliver finance. In Rwanda, the Banque Rwandaise de Développement/Development Bank of Rwanda manages the FONERWA credit line for private sector investment in LCRD. In Ethiopia, microfinance institutions and the Development Bank of Ethiopia deliver finance for investment in projects that are consistent with CRGE aims. Commercial financial institutions – for instance commercial banks such as the Nib International Bank in Ethiopia and the Bank of Kigali in Rwanda – may also provide support for private sector investments.

Finally, the national climate change funds – the CRGE facility in Ethiopia and FONERWA in Rwanda – play a key role in shaping the climate finance landscape in these countries. The institutional arrangements for running FONERWA include a management committee operating at the highest level of government. This has overall responsibility for monitoring and directing the fund's activities: it approves budgets and work plans, and ultimately makes the funding decisions. There is also a technical committee, supported by a team from the finance ministry. The latter has the role of ensuring that FONERWA activities do not duplicate activities already included in annual plans, and that they are aligned with the priorities of Rwanda's National Development Plan. The staff of the management and technical committees is provided by central and sectoral ministries.

The CRGE Facility has a management team, a technical team, 'implementing entities' and 'executing entities'. The management team is responsible for the financial management of the fund, and is housed in Ethiopia's MOFED. The technical team is responsible for coordinating the CRGE planning process and is housed in the Ministry of Environment Protection and Forestry. The implementing entities are sector ministries responsible for identifying CRGE investment priorities, while the executing entities are public, private and civil society organisations responsible for managing spending on CRGE investments. It should be noted that in 2016 Ethiopia's government was considering how best to align the CRGE Facility's institutional structures with existing public finance management structures.

The various actors we have described work with particular financial intermediaries, instruments and planning systems to deliver finance for LCRD. In the next section we explore the different approaches they take to meeting the challenges of this task.

Actors' policy choices

Countries need scaled-up, long-term and appropriate finance to meet the cross-sectoral, programmatic and long-term needs of LCRD investments. Various actors – international and national – are exploring policy options for national-level climate finance in order to meet the ambitions of the LCRD agenda. Depending on their policy mandate and capacity, policymakers target specific sources of finance for investment in priority areas, and work with particular financial intermediaries, instruments and planning systems. Based on our analysis of interviews with government and donor stakeholders in Ethiopia and Rwanda, we have identified two broad approaches to mobilising and delivering LCRD finance: a single-channel modality and a multi-channel modality.

The single-channel modality

A single-channel modality involves policy actors targeting a single source of finance, and then channelling it to end users via a single financial intermediary, instrument and financial planning system. It reflects a more 'traditional' concept of the funding process, and is usually best suited to accessing specific sources of finance for a specific project- or sector-level investment.

Actors that tend to favour this approach include those in multilateral banks, multilateral and bilateral agencies, and national government agencies. Multilateral and bilateral agencies mostly target international sources of public finance, while national government agencies additionally target national public finance. These actors then tend to use financial instruments that minimise the financial risks for the end user, such as grants, guarantees, insurance and concessional loans. They often channel finance through public sector financial intermediaries such as multilateral development banks and agencies, national agencies and national development banks, and rely on public systems to manage the flow of climate finance.

As an example, Ethiopia's Ministry of Water, Irrigation and Energy channels funds for renewable energy investments through public agencies such as EEPCO (the Ethiopian Electric Power Corporation) using single grant-based instruments. Similarly, the IFC is an international multilateral actor mandated to channel funds through private sector agencies using a single loan-based modality.

The single-channel modality was previously the one most commonly used, and has funded many development finance interventions. Depending on the nature of the investment it can have significant advantages. For example, for a specific public project a simple single channel approach in which a public agency provides grants for short-term infrastructure goods can be effective; in particular it will make it easy to demonstrate outcomes.

The multi-channel modality

LCRD investments typically require finance on a large scale, over a relatively long period of time, and on appropriate terms, according to the programme stage. This may require a range of financial entities and instruments. The cross-sectoral nature of LCRD projects also means that a variety of implementing entities are likely to be involved.

These characteristics mean that a multi-channel modality is likely to be a suitable option. This involves policy actors targeting multiple sources of finance and channelling funds to diverse users and initiatives, employing a range of financial intermediaries, instruments and planning systems. Actors who tend to favour this type of approach include international and national climate funds and national development finance institutions. They target a range of international and national sources of public and private finance and are able to deliver resources to a range of public and private sector investors. They use a variety of financial intermediaries, instruments and financial planning systems to access and deliver finance.

A multi-finance approach to resource mobilisation and delivery is reflected in the official policy narratives of both Ethiopia and Rwanda, as articulated in their national development climate change policies and in operational manuals linked to the climate change funds. In both countries policy is shaped by key guiding principles aimed at delivering appropriate finance for investment in LCRD. One such principle is that of 'leveraging' or 'catalysing' finance from different sources – national and international, public and private – in order to mobilise finance on the scale required. Other guiding principles include national ownership and programmatic delivery, which are aimed at achieving efficient and effective financial flows.

A key characteristic of the multi-channel modality seen in our focus countries is that they mix and match intermediaries to diversify the options for accessing and channelling climate finance for inclusive investment in LCRD. Ethiopia's national climate change fund, the CRGE Facility, is the primary intermediary for CRGE investments. It is designed to pool funds from multiple international and national sources, thereby mobilising resources efficiently. So far it has successfully accessed bilateral sources of climate finance, and it is in the process of applying for accreditation for the Adaptation Fund and the GCF under UNFCCC in order to access multilateral sources directly. It is also expected to work with other financial intermediaries; for example, it will work with national financial institutions to disburse public and private finance for private sector investments. The CRGE Facility enables Ethiopia to manage climate funds within a single coherent system that allows investors to determine best practice. This 'programmatic' approach aims to minimise transaction costs and reduce the fragmentation and duplication associated with funding unconnected projects.

Rwanda's national climate change and environment fund, FONERWA, is designed to evolve as different sources of finance and new areas of investment become viable. In the short to medium term the country's Ministry of Natural Resources will manage the fund, while the Development Bank of Rwanda will manage a credit facility to incentivise private sector investment. In the long term, if investment in low carbon, climate resilient development becomes commercially viable, FONERWA has the scope to be managed as a venture capital fund.

A second key characteristic of the multi-channel modality is the sequenced deployment of economic and financial instruments to incentivise scaled-up public and private sector investment. We use the term 'economic instrument' here to refer to any framework, including a policy or regulatory one, that influences producers' and consumers' behaviour by effecting price changes. An example of this is the Ethiopian government's use of power purchase agreements and feed-in tariffs to encourage investment in renewable energy.

A financial instrument, meanwhile, is any contract that gives one entity a financial asset and another a financial liability. Financial instruments that incentivise LCRD investments include risk management instruments such as guarantees and insurance, grants and concessional loans, and capital instruments such as equity and debt finance. Different instruments suit different investment needs: risk management instruments enable investors to make high-risk investments; grants are effective in supporting investments in climate resilience; and capital instruments are effective once LCRD investments are commercially viable.

The Ethiopian government is planning to provide a range of financial instruments through the CRGE Facility. These include grants, concessional loans and results-based payments. Similarly, the Rwandan government is planning to deploy financial instruments in a phased approach to support the evolving financial needs of LCRD investments. Short-term financial instruments (operating for up to a year) will include in-kind support such as technical assistance, grants to support public sector investors and performance-based grants to support private sector investors. Medium-term financial instruments (operating for two to five years) will include guarantees and low-interest concessional loans. Long-term financial instruments (operating for more than five years) will include equity investments, subject to FONERWA's performance and private sector demand.

Both countries plan to use a range of financial planning systems to manage the flow of climate finance. In Rwanda, the government has used budget and planning systems to leverage greater synergy between investments. Funds disbursed by FONERWA are incorporated into the annual budget allocation of government ministries and public sector agencies to encourage integrated investment and avoid duplication and fragmentation. The reporting systems of the Development Bank of Rwanda are used to account for FONERWA funds disbursed to the private sector. In time the government plans to use private sector management systems (venture capital funds) to finance private sector investment in LCRD. Ethiopia is currently assessing how best to manage climate finance using public finance management systems, including its annual budget and planning cycles.

Both countries have developed policies to establish their climate change funds and operational manuals to guide the flow of resources channelled through them. These outline institutional structures support synergies between different sources of finance and different investment portfolios.

In this section we have outlined the emergence of a multi-channel modality alongside a more traditional single-channel approach to resource mobilisation and delivery of finance for investment in LCRD in Ethiopia and Rwanda. This can be seen as a response to a changing financial landscape and investment needs. Early climate change investments focused on piloting projects to address adaptation priorities, as outlined in NAPAs and by the CIF programmes. These investments largely relied on multilateral and bilateral sources of finance that were delivered through multilateral agencies using grants and concessional loans.

Current LCRD investment in Ethiopia and Rwanda focuses on a range of short-, medium- and long-term initiatives that are designed to create an enabling environment for future investment, mainstream LCRD initiatives into existing development planning and incentivise investment in new LCRD business models. Public and private sector investors at international, national and local level are all expected to support these initiatives. A multi-channel modality is better suited to accessing finance from a range of sources

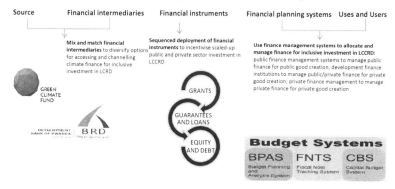

Figure 5.1 Policy options for mobilising and delivering finance for inclusive investment in LCRD
Source: Kaur et al. (2016), available at http://pubs.iied.org/10138IIED, © IIED.

and to meeting the needs of public and private sector investors. Both countries aim to use a range of financial intermediaries, instruments and planning systems to incentivise scaled up and long-term finance for public and private sector investment in LCRD. The case study in Box 5.1 illustrates how a mix of financial intermediaries and instruments can be effective in leveraging and delivering additional finance.

Box 5.1 Ethiopia's Coffee Initiative

The Coffee Initiative was initially supported by a US $10 million grant from the Bill and Melinda Gates Foundation to provide business solutions to coffee cooperatives. These business solutions were aimed at boosting the incomes of smallholder coffee farmers by improving the viability, governance and quality management systems of 'wet mills' used in coffee production; increasing productivity through field-based agronomy training; and optimising the overall farm-to-market value chain.

The financial intermediaries responsible for channelling the finance to end users included TechnoServe, the IFC and the Nib International Bank. In 2010 TechnoServe helped to establish a new relationship between the IFC and the Nib International Bank designed to unlock the substantial capital required for investment in coffee production and to enable smallholders to access finance without collateral.

This arrangement was facilitated by the use of three financial instruments. TechnoServe provided a guarantee of US $10 million to the IFC. This enabled the IFC to establish a $10 million risk-sharing facility with the Nib International Bank, which in turn was able to provide a revolving loan facility to 62 coffee cooperatives, representing approximately 47,000 farmers. The facility offered up to $250,000 per cooperative, disbursed against cash flow requirements and collateralised by coffee stocks.

> In terms of the initiative's outcomes, the financial instruments operated to provide unbanked cooperatives with access to credit, which in turn has unlocked additional investment in coffee production from private banks and the cooperatives themselves.
>
> Participating cooperatives have been able to export over 2 million pounds of high-quality coffee to 12 international buyers in Europe and the United States, at prices that were on average 40 per cent higher than those of low-quality unwashed coffee. By the end of the first phase of the programme, the cooperatives had earned US $11 million in revenue and the finance provided had directly contributed to the creation of more than 1,000 wet mill jobs.
>
> In addition to the loan guarantees provided by the IFC, the Coffee Initiative secured guarantees from Rabobank and from Falcon Commodities, a major coffee buyer. As part of these guarantees, local Ethiopian banks increased their share of credit risk from 0–35 per cent when lending to Coffee Initiative clients. There is an expectation that by gradually encouraging local banks to lend to the coffee sector the financial services available to coffee cooperatives will continue to expand.
>
> Adapted from TechnoServe (2013a, 2013b, 2016).

It is also worth noting that these two approaches do not exist in isolation and do not assume a fixed form. Actors may use either model or a combination of both, depending on the context. For example, historically some environment ministries primarily used grant-based instruments to channel finance for specific projects. However, environmental agencies have adapted their roles and approaches over time in order to meet the cross-sectoral, long-term needs of climate interventions, using a wider range of instruments and channels as needed. In sum, roles and approaches are evolving and will continue to do so as optimal LCRD outcomes are sought.

Actors, policy networks and incentives providing support for specific financing approaches

This book uses the concept of discourse coalitions and policy networks to explore the convergence of ideas and policy narratives around particular actions (see Chapter 1). In this chapter we also make a distinction between 'discourse coalitions' and 'policy networks', which deliberately engage in joint action to promote particular policy options (see Marsh and Rhodes 1992). Our focus is on examining the extent to which policy networks support single- or multi-channel modalities.

Figure 5.2 illustrates the coalitions emerging around the different approaches to LCRD finance. Actors clustering around either approach do so due to the incentives provided by their policy mandates and capacity constraints and strengths. Multilateral banks and national agencies tend to favour a single-channel modality, due to path dependence and experience of working with

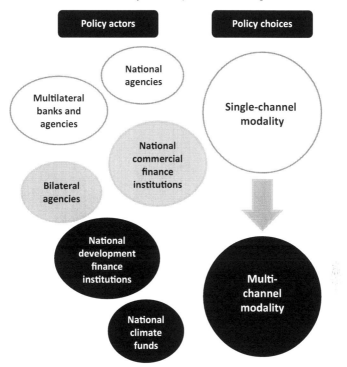

Figure 5.2 Coalitions supporting approaches to financing inclusive investment in LCRD
Source: Kaur et al. (2016), available at http://pubs.iied.org/10138IIED, © IIED.

particular financial intermediaries, instruments and planning systems. Bilateral agencies and national commercial banks tend to target single sources of finance, but are able to use a range of financial intermediaries and instruments to deliver it. National development finance institutions and climate funds are able to target multiple sources of finance and use a range of financial intermediaries, instruments and planning systems.

We now go on to consider the extent to which the two financing modalities are supported by policy networks, given that each has its benefits and limitations for mobilising scaled-up and long-term finance for LCRD.

Actors with shared ideas do not necessarily see themselves as part of a network; they may not work together and indeed may not even be aware of other actors with similar views. In the two case study countries policy networks based on joint action have yet to emerge. Various actors tend to align with a particular approach according to their interests and practice, but there is a lack of communication and coordination among them. The formation of policy networks could be important to achieving the necessary scale and type of LCRD finance, however. For instance, national and international policy actors could benefit from sharing their experiences of using particular financial intermediaries, instruments and planning systems. Such an exchange might enable international financial intermediaries like the GCF to put in place incentives that enhance the capacity of national financial intermediaries. At national level, greater communication between actors supporting single and multi-channel approaches could enhance learning and coordination, and enable policymakers to leverage improved financial outcomes. Cross-learning and coordination might also result in the evolution of networks over time.

Conclusions

Mobilising and delivering finance for inclusive investment in LCRD is a new and complex policy area. It is clear that there is a mismatch between national LCRD investment priorities and the scale and type of international and national finance available. Our study shows that policymakers in Ethiopia and Rwanda are identifying effective policy options for mobilising and delivering climate finance for LCRD. The results of our analysis can be summarised in three main findings:

- There are two main approaches to mobilising and delivering LCRD finance: a more traditional single-channel modality and a newer multi-channel modality. The former uses a single source of finance, which is then channelled to specific users via single intermediaries, instruments and systems. A multi-finance modality involves a range of sources of finance, and uses a range of instruments and entities to channel this finance.
- A single-channel modality is best suited to addressing short-term, project-specific climate change priorities. However, LCRD investment needs are generally longer term, programmatic and cross-sectoral. A multi-channel modality is better suited to meeting these needs, as it can provide finance on an adequate scale and channel it via cross-sectoral institutions and mechanisms.
- The two financing modalities are supported by different discourse coalitions. Multilateral and bilateral agencies and national agencies, traditionally tied through aid partnerships, typically follow a single-channel modality to deliver project-specific outcomes, as do national agencies that use public sources of finance. Meanwhile, actors such as multilateral climate funds and national development finance institutions, including development banks, are often supporters of a multi-channel approach.
- Factors that incentivise actors to follow a particular approach to finance include path dependence, the existence of pre-existing relationships with particular actors and experience of using specific instruments and processes. Institutional mandates may also be a factor. This can determine

the instruments used, for example line ministries may only be able to provide grant funding while some multilateral development banks are restricted to offering loans. The type of project may also be important; for example, a funding agency's mandate restricts it to financing short-term projects.

The nature of the LCRD agenda incentivises policy actors to establish entities that use a multi-channel modality, with mandates to mobilise and blend finance from different sources in order to fund long-term, programmatic investments requiring cross-sectoral cooperation.

These findings suggest that an understanding of actors, policy networks and policy approaches, including the factors that incentivise these approaches, can help to improve coordination among actors implementing LCRD finance. Shared understanding and synergies between policy options will result in better resource mobilisation and delivery of climate finance for cross-sectoral LCRD investment (UNEP 2015; Schmitz and Scoones 2015). Levers for influencing LCRD finance can be used to ensure that:

- intermediaries are positioned to deliver funding at the right scale and to the right people;
- financial management systems are designed so as to channel the flow of funds in ways that are effective in meeting LCRD priorities;
- financial instruments – from grants to equity and debt finance – are used to incentivise investment, with different types and combinations deployed according to context.

We have found that discourse coalitions are emerging to support the different policy options. They have yet to evolve into policy networks, but if they do, such networks can shape the financial landscape and help to establish effective policies for financing inclusive investment in LCRD (UNEP 2015). They can also improve coordination between actors, thus facilitating better synergy and alignment of policy options.

It follows that policymakers can support coalitions involved in national climate finance by structuring incentives appropriately to develop policy networks. These incentives might include:

- Policy and regulatory incentives that provide a mandate for better links between financial intermediaries. For instance, in Rwanda the FONERWA operational manual provides a mandate for the country's climate change fund to work with the national development bank.
- Policy and regulatory incentives that help to align private sector interests with public goals and to ensure financial inclusion.
- Capacity incentives that strengthen the ability of financial intermediaries to deliver coordinated financial support for investment in LCRD.

- Financial incentives that encourage traditional and new actors in the climate finance landscape to work with each other.
- Knowledge incentives such as peer learning and experience sharing platforms to help actors develop a better understanding of how single- and multi-finance approaches work to deliver finance for investment in LCRD.

References

Boyle, J. (2013) *Exploring trends in low-carbon, climate-resilient development*. Manitoba: IISD. Available at http://www.iisd.org/pdf/2013/exploring_trends_low_climate.pdf (accessed 23 March 2016).

Buchner, B., Herve-Mignucci, M., Trabacchi, C., Wilkinson, J., Stadelmann, M., Boyd, R., Mazza, F., Falconer, A. and Micale, V. (2013) *The global landscape of climate finance 2013*. Climate Policy Initiative Report. Venice: CPI. Available at http://climatepolicyinitiative.org/wp-content/uploads/2013/10/The-Global-Landscape-of-Climate-Finance-2013.pdf (accessed 25 January 2016).

Eshetu, Z., Simane, B., Tebeje, G., Negatu, W., Amsalu, A., Berhanu, A., Bird, N., Welham, B. and Trujillo, N. C. (2014) *Climate finance in Ethiopia*. London: ODI. Available at http://www.odi.org/publications/8203-climate-finance-ethiopia (accessed 23 March 2016).

Fikreyesus, D., Kaur, N., Kallore, M. E. Ayalew, L. T. (2014) *Public policy responses for a climate resilient green economy in Ethiopia*. IIED Research Report. London: IIED. Available at http://pubs.iied.org/10066IIED.html (accessed 23 January 2016).

Fisher, S., Fikreyesus, D., Islam, N., Kallore, M., Kaur, N., Shamsuddhoa, Md., Nash, E., Rai, N., Tesfaye, L. and Rwirihira, J. (2014) *Bringing together the low-carbon and climate resilience agendas: Bangladesh, Ethiopia, Rwanda*. IIED Working Paper. London: IIED. Available at: http://pubs.iied.org/10099IIED [Accessed 23 January 2016].

Government of Ethiopia (2011). *Ethiopia's climate resilient green economy: Green Economy Strategy*. Addis Ababa: Federal Democratic Republic of Ethiopia.

Government of Rwanda (2013) *Economic Development and Poverty Reduction Strategy II: 2013–2018*. Kigali: Government of Rwanda. Available at http://www.rsb.gov.rw/~rbs/fileadmin/user_upload/files/EDPRS_2_Abridged_Version.pdf (accessed 23 March 2016).

Kaur, N., Tesfaye, L. and Mamuye, S. (2016) *Financing inclusive investment in low-carbon resilient development: the role of national financial institutions in Ethiopia's Climate Resilient Green Economy*. IIED Country Report. Available at http://pubs.iied.org/10138IIED.html (accessed 29 March 2016).

Kaur, N. Rwirahira, J. Fikreyesus, D., Tesfaye, L. and Mamuye, S. (2016) *Financing inclusive investment in low-carbon climate resilient development: the national climate finance landscape in Ethiopia and Rwanda*. IIED Working Paper. London: IIED. Available at http://pubs.iied.org/10110IIED.html (accessed 23 March 2016).

Marsh, D. and Rhodes, R. A. W. (eds) (1992) *Policy networks in British government*. Oxford: Clarendon Press.

Ministry of Finance and Economic Development (2010) *Growth and Transformation Plan (GTP) 2010/11–2014/15. Vol. 1: main text*. Addis Ababa: Government of Ethiopia.

OECD-DAC (2013) Development finance statistics database. Available at http://www.oecd.org/development/stats (accessed 1 December 2015).

Rai, N., Iqbal, A., Zareen, A., Mahmood, T., Muzammil, M., Huq, S. and Elahi, N. (2015) *Financing inclusive low-carbon resilient development: role of Central Bank of Bangladesh and Infrastructure Development Company Ltd*. IIED Country Report. London: IIED. Available at http://pubs.iied.org/10139IIED.html?b=d (accessed 16 December 2015).

REMA (2010) *Operationalisation of National Fund for Environment (FONERWA) in Rwanda*. Kigali: REMA.

REMA (2011) *National Strategy on Climate Change and Low Carbon Development for Rwanda: baseline report*. Oxford: Smith School of Enterprise and the Environment, University of Oxford. Available at http://www.rema.gov.rw/fileadmin/templates/Documents/rema_doc/CC%20depart/final-baseline-report-rwanda.pdf (accessed 23 March 2016).

Schmitz, H. and Scoones, I. (2015) *Accelerating sustainability: why political economy matters*. IDS Evidence Report 152. Brighton: IDS. Available at http://www.ids.ac.uk/publication/accelerating-sustainability-why-political-economy-matters (accessed 23 March 2016).

Steinbach, D., Acharya, S., Bhushal, R.P., Chhetri, R.P., Paudel, B. and Shrestha, K. (2015) *Financing inclusive low-carbon resilient development: the role of the Alternative Energy Promotion Centre in Nepal*. IIED Country Report. London: IIED. Available at http://pubs.iied.org/10140IIED.html (accessed 23 March 2016).

TechnoServe (2013a) *The Coffee Initiative: phase one final report, 2008 to 2011*. Washington, DC: TechnoServe. Available at http://www.technoserve.org/files/downloads/coffee-initiative-phase-one-final-report.pdf (accessed 23 March 2016).

TechnoServe (2013b) *Lessons learned: the Coffee Initiative, 2008 to 2011*. Washington, DC: TechnoServe. Available at http://www.technoserve.org/files/downloads/coffee-initiative-lessons-learned.pdf (accessed 23 March 2016).

TechnoServe (2016) *Where we work: Ethiopia*. Available at http://www.technoserve.org/our-work/where-we-work/country/ethiopia#_resources (accessed 23 March 2016).

United Nations Environment Programme (UNEP) (2015) *The financial system we need: aligning the financial system with sustainable development. The UNEP Inquiry report*. Geneva: UNEP. Available at http://web.unep.org/inquiry/publications (accessed 23 March 2016).

6 Incentives and opportunities for local energy finance

Neha Rai and Dave Steinbach

Introduction

Investment in LCRD is intended to enable citizens, households and economies to achieve development gains as well as to mitigate and adapt to the effects of climate change. The lower-income groups that form the 'base of the pyramid' will bear the brunt of the worsening impacts of climate change, and LCRD policies need to include these groups in order to improve their resilience to climate-related shocks, their access to services such as health and education, and their opportunities for building sustainable livelihoods.

LCRD investment may be in anything from improved crop varieties to water management technologies, from solar energy to climate resilient infrastructure and business development. As countries move to implement their low carbon action plans, it is important to understand how these plans are being delivered on the ground and to ask how we can ensure that climate-related interventions benefit those who need them most.

In this chapter we use a political economy framework to analyse actors and incentives that catalyse decentralised energy projects in LDCs. The analysis is based on IIED research into projects in Bangladesh, Ethiopia, Nepal and Rwanda involving technologies such as Solar Home Systems, Solar Irrigation Pumps, biogas, and mini- and micro-grids in rural areas (Steinbach et al. 2015a; Rai et al. 2015a; Rwirahira and Fisher 2015; Kaur et al. 2016). We use this analysis to explore how communities and markets can invest in and benefit from local low carbon initiatives. We provide an overview of the modalities used – the financial instruments, intermediaries and systems – and consider their effectiveness. We look at what incentivises intermediaries to invest in renewable energy technologies in low-income communities and what incentivises members of those communities to purchase those technologies. Finally, we draw on this evidence to suggest ways of structuring delivery mechanisms to encourage pro-poor LCRD.

Our analysis leads us to make three arguments:

- First, decentralised energy can deliver co-benefits to households at the bottom of the pyramid, but for this to happen these households *need*

access to finance, and their specific financial needs remain unfulfilled by mainstream markets.
- Second, it is possible to design financial mechanisms that reach out to low-income communities. Some *innovative examples* of decentralised energy targeting the base of the pyramid show that appropriately tailored financial intermediaries and instruments can give the poor access to finance on the right terms, but these examples are limited.
- Third, with the right *incentives* in place for both investors and communities, low-income populations can access finance to purchase alternative energy options and derive local co-benefits from LCRD.

We discuss these points in turn, beginning with the financial needs of low-income communities and the extent to which they are being met.

Energy access and energy finance in low-income communities

Providing local communities with adequate access to energy can support a range of development objectives, including improved public health, education, water and sanitation, and gender equality, among many others. Health benefits, for example, can result from reduced indoor air pollution; and women can be empowered through having more time for productive activities or education (Pueyo et al. 2013; Walters et al. 2015). Access to electricity spurs economic development through, for example, reducing household energy expenditure, improving small business opportunities and increasing connectivity and communication. And if this energy comes from cleaner, renewable sources it also has the benefit of reducing carbon emissions. In this way, by delivering development co-benefits at the same time as building climate resilience and mitigating greenhouse gas emissions, LCRD can represent a triple win.

However, nearly 4.5 billion people worldwide live below the US $2-a-day poverty line, at the bottom of the income pyramid, with limited ability to pay for essential public goods such as electricity (Prahalad and Hart 2004). Eighty per cent of the world's poor live without access to electricity in rural areas of Asia and Africa (International Energy Agency 2012). There is a disjunction between the financial needs of these low-income communities and the financial products available to them for investment in LCRD. Investors are wary of becoming involved in such markets because of the risk involved and also a lack of expertise in this area. So that better approaches can be designed, there is a need to understand the underlying political economy here, in particular what incentivises (and disincentivises) both financial actors and low-income end users to invest in LCRD.

In our case study countries, two-fifths of Bangladesh's population is 'off-grid', or not connected to any form of central or mini-grid (Rai et al. 2015a). Around 70 per cent of Nepal's population has access to electricity, but this is mostly in urban centres. Poor transmission and low generation capacity create a bottleneck in supply to more isolated, rural areas. A rural–urban divide can also be seen in Ethiopia, where of the 83 per cent of the population living in rural areas only 4 per cent have access to electricity (Central Statistical Agency 2012). Rwanda, meanwhile, consumes an average of only 42 kilowatt hours per person per day in electricity, compared to a sub-Saharan average of 478 kilowatt hours per person per day (Government of Rwanda 2013; Rwirahira and Fisher 2015).

Decentralised energy is an effective way of providing rural populations with access to electricity, particularly in those areas where grid connection is difficult. However, utility companies tend to invest in grid connections in urban centres; they find it difficult and costly to reach more isolated rural communities, where the scattered nature of the population means that they cannot achieve economies of scale and transaction costs are high. Meanwhile, the cost of technology, appliances and connection, and the difficulty of getting finance discourage rural households from attempting to access electricity (Watson et al. 2011; Pueyo et al. 2013).

At the same time, the current international and national financial landscape is failing to meet the needs of low-income countries and communities in terms of funding energy provision. According to the Overseas Development Institute's Climate Funds Update database, at 2015 only approximately 3 per cent of the total approved public climate finance was for decentralised energy projects. Instead, the primary recipients of energy sector funding were energy efficiency and large-scale utility projects in middle-income countries, predominantly financed by concessional loans from multilateral development banks (Climate Funds Update 2015). This suggests that the banks' preference for using loans and the perceived 'bankability' of projects are the factors that are driving energy investment choices, and that the effect thereof has been to exclude decentralised projects in low-income countries.

We suggest that the main reason communities at the bottom of the income pyramid have insufficient access to electricity is that they have specific financial needs, and these needs are not being met. It is not just a matter of improving the scale of finance to the poor, but also of providing it on appropriate terms – of making it accessible and predictable through using instruments and intermediaries that cater specifically to this group.

The needs of low-income communities also go beyond simple availability of credit. The different actors involved – not only households and communities but also the suppliers, enterprises and investors, the SMEs and NGOs that serve them – all have particular needs and require finance for different purposes:

- Low-income households need upfront capital as well as affordable finance in order to switch from low-quality, unsustainable energy sources such as kerosene to renewable technologies. Finance needs to be on terms appropriate to smaller-scale projects, and it needs to be cost-effective.
- Investors in low-income markets look for proven business models, a clear picture of the risks and returns, and indications that risks are adequately managed. They need assurances that their loans to low-income off-grid

communities will be paid off. They also need adequate returns on these small-scale investments, which have high transaction costs. Clear policy signals are needed for assured market development (World Economic Forum 2012; Wilson et al. 2014).
- Manufacturers and suppliers that cater to these low-income markets require finance to purchase assets, for working capital and for their start-up and development period, until they can make a profit (Ashden and Christian Aid 2014).

Financial instruments and incentives need to be designed in ways that take into account these needs, as well as local capacities (UNDP 2011) and the relatively small scale of projects and income generated for private investors (Rai et al. 2015a). Similarly, appropriate financial intermediaries are needed, with the ability to channel funds to the local level, and to both end users and investors (Glemarec 2012).

Many LDCs are building new administrative architectures – institutions, intermediaries, instruments and planning systems – to deliver LCRD projects (Kaur et al. 2014; Rai et al. 2015a). However, empirical studies have demonstrated that such initiatives often fall short of their intended targets and fail to make a real difference to those who are most vulnerable to climate change (Wilson et al. 2014; Sharma et al. 2015). This mirrors the situation in international development finance, very little of which actually trickles down to those it is meant to reach. Furthermore, the people most at risk seldom have a say in how projects are prioritised or who is involved (Steele et al. 2015b).

A key issue is that a lot of LCRD finance gets absorbed by institutions operating at national scale before it can reach poorer communities at local scale (Christensen et al. 2012). So although these institutions are often essential to a country's ability to unlock finance, in order to deliver this finance to the groups that need it most there needs to be greater integration of financial channels and more targeted and cost-effective financial instruments. This in turn means improving incentives and reducing disincentives (Rai et al. 2015a).

How innovative delivery models are helping low-income communities to gain access to energy

Turning to our second argument, then, in this section we look at how such tailored delivery models can provide energy access for the poor on the right terms, examining case studies of innovative decentralised renewable energy projects in Bangladesh, Ethiopia, Nepal and Rwanda.

Decentralised energy access projects in Bangladesh, Ethiopia, Nepal and Rwanda

Table 6.1 provides a summary of decentralised renewable energy projects in our four case study countries. We can see that these countries are using

110 *Incentives for financing decentralised energy*

Table 6.1 Public-private investment in decentralised renewable energy in four LDCs

Country	National agencies and programmes	Renewable energy projects
Bangladesh	Central Bank of Bangladesh Infrastructure Development Company Ltd (IDCOL)	Solar Home Systems and Solar Irrigation Pumps
Ethiopia	Development Bank of Ethiopia Market Development for Renewable Energy and Energy Efficient Products programme	Off-grid renewable energy and energy efficient technologies, e.g. solar lanterns, biogas digesters
Rwanda	Development Bank of Rwanda/ Banque Rwandaise de Développement FONERWA – national environment and climate change basket fund	Renewable energy for industry, e.g. Rwanda Mountain Tea Corporation National Domestic Biogas Programme
Nepal	Alternative Energy Promotion Centre (AEPC) Central Renewable Energy Fund (CREF) National Rural Renewable Energy Programme (NRREP)	Small-scale renewable energy such as mini- and micro-hydro, solar, cooking stoves

public-private investment models to meet energy needs in different ways, thus promoting investors' involvement and helping households to access finance.

In Bangladesh, LCRD policies and programmes are aimed principally at widening access to energy. In one of the world's largest off-grid electrification projects, the government of Bangladesh is encouraging households to purchase Solar Home Systems. Nearly four million homes now use one of these systems, enabling them to avoid some of the cost and many of the inconveniences of diesel or kerosene, and to gain a variety of health and educational benefits. A more recent initiative is the Solar Irrigation Pumps scheme, which is targeted at farmers. These pumps make electricity available to farms at the right time of year, reducing the amount of labour and fertiliser needed and hence reducing cultivation costs. They also increase productivity by enabling triple cropping and by freeing up time for tasks other than irrigation. The two key entities involved in implementing these schemes are the country's central bank and its Infrastructure Development Company Ltd (IDCOL), which work with intermediaries such as NGOs, microfinance institutions and local banks to mobilise and deliver finance.

Energy access, quality of energy supply and productive energy use are the priorities outlined in the national LCRD plans of both Rwanda and Ethiopia. The Development Bank of Ethiopia works in partnership with microfinance institutions to deliver the Market Development for Renewable Energy and Energy Efficient Products programme. This supports households, private sector enterprises and SMEs to invest in off-grid renewables and energy efficient technologies, including solar lanterns and biogas digesters. These bring a variety of benefits, including reduced indoor pollution, promotion of education among both adults and children, and improved communications (by increasing access to radios and mobile phones). Economic benefits take the form of household savings – due to reduced expenditure on charcoal, fuel woods or kerosene, for example – and increased income, which is the consequence of improved business opportunities and productivity. For instance, there is increased use of fertiliser, which is a by-product of biogas digesters.

Rwanda is also implementing projects through its national development bank and FONERWA. The latter helps private companies to provide renewable energy to industry. While this approach helps to improve Rwanda's energy security, benefits to communities are indirect and may be the result only of private sector corporate social responsibility activities. In contrast, the National Domestic Biogas Programme, which promotes cooking with biogas rather than with firewood, is aimed directly at households.

In Nepal, the AEPC, a semi-governmental agency, runs an NRREP (launched in 2011) which aims to bring together renewable energy projects within a single programme. The AEPC has also established a CREF to channel finance for small-scale renewable energy, particularly in rural areas, via commercial banks and microfinance institutions.

Delivery models

The various programmes in operation across the four case study countries show that there are two main financing channels for decentralised energy investments: special purpose agencies, and central and national development banks. Both use specific delivery models to incentivise the flow of finance to low-income markets, as well as a combination of financial intermediaries and instruments to promote inclusive investment.

Special purpose agencies

'Special purpose agencies' in this context are semi-autonomous legal entities set up to meet specific objectives, and with the ability to pool funding and expertise. The intention is to accelerate the flow of funds from a range of donors and government sources to financial intermediaries and end users. These agencies are able on the one hand to draw down large-scale finance, and on the other to provide a basket of services and a range of instruments appropriate to the particular needs of low-income groups. This enables the poor to access finance in the amounts and on the terms they need, with flexible, affordable capital made available over the longer term.

Bangladesh's IDCOL and Nepal's AEPC are two examples of this type of agency. Both have been set up specifically to meet small-scale renewable energy needs in decentralised rural communities (see Box 6.1 for more details). Both use an incentive-based, phased subsidy approach. In the early stages subsidies are provided to financiers and developers to promote the uptake of renewable energy technologies, but in an environment that sets up a shift towards credit financing in the longer term. This environment is the result of a 'one-stop shop' model, whereby the role of the special purpose agency is not limited to finance, but also includes a variety of services aimed at helping to create markets and delivery networks. These provide access to capital, quality assurance, after-sales services, training and institutional support (Rai et al. 2015b).

Box 6.1 Special purpose agencies promoting renewable energy provision in low-income off-grid communities

Bangladesh: IDCOL

IDCOL is a government-owned financial intermediary set up to encourage the provision of private finance for renewable energy projects in Bangladesh, particularly in areas where grid connection is difficult; the direct beneficiaries of IDCOL projects are therefore rural households and communities. It provides financial support for renewable energy technologies such as Solar Home Systems, Solar Irrigation Pumps, domestic biogas, solar mini-grids, solar-powered telecoms, a biogas-based electricity project, a biomass gasification project and improved cooking stoves (Islam 2014).

In the case of the Solar Home Systems programme, IDCOL provides households with capital buy-down grants to cover a down payment of US $17 on the system, which costs $193 in total, while the households pay around $20 out of their pocket. The buyer takes out a three-year loan from the partner organisation to cover the remaining $156, typically at an interest rate of 15–20 per cent. Once the system has been installed, 70–80 per cent of this loan is refinanced by IDCOL to the partner organisation at a rate of around 6–9 per cent (Asaduzzaman et al. 2013; Rai et al. 2015a). This arrangement enables households to adopt renewable energy solutions by providing them with both grant assistance and access to credit. The involvement of partner organisations, meanwhile, is incentivised by concessional loans from IDCOL that enable them to on-lend to households at a profit, as well as by institutional development grants also provided by IDCOL.

In the case of Solar Irrigation Pumps, users are provided with direct grants and loans as primary owners of the system. The grant component is relatively high, at up to 50 per cent of the total project cost.

Beyond financial support, IDCOL also offers a basket of services to support the delivery of energy access projects. These include support for capacity building, training for partner organisations and quality assurance.

Nepal: the AEPC and the CREF

The AEPC is a non-banking development finance institution established to support renewable energy projects, with funding coming from the Nepalese government and development partners. The AEPC initially provided project funding mainly through subsidies; however, recently a CREF has been set up to facilitate a shift to credit-based funding. This fund is now embedded within the commercial Global IME Bank. The bank provides subsidies to qualified renewable energy installers and loans through commercial partner banks to suppliers, manufacturers, installers and communities.

In 2012 the AEPC launched the US $170 million NRREP, which is funded by the government of Nepal and bilateral and multilateral development partners. The programme funds the installation of renewable energy technologies such as micro- or mini-hydropower, Solar Home Systems, institutional solar power systems and improved cooking stoves. This funding is provided to households and communities in the form of grants and credit: 40 per cent of the cost of installation is covered by a grant from the Global IME Bank, 40 per cent by an NRREP loan and the remaining 20 per cent through co-finance from households, districts and village-level committees (see Steinbach et al. 2015a and Rai et al. 2015a for further details).

Central banks and national development banks

In addition to its creation of a special purpose agency for LCRD, Bangladesh is also using its central bank to incentivise investment in renewable energy through commercial banks, microfinance institutions and NGOs. Ethiopia and Rwanda are using their respective national development banks for the same purpose (see Box 6.2 for more details). These are domestic government institutions with a specific policy mandate to provide long-term finance to more risky sectors not typically served by commercial banks. They have been used for some time to channel development finance; increasingly they are also being used to channel climate finance.

Box 6.2 Central and national development bank

The Central Bank of Bangladesh

The Central Bank of Bangladesh was the first central bank in the world to dedicate resources to green projects. In 2005 it set up a refinancing scheme advising commercial banks on finance for green energy, including Solar Home Systems (also funded by IDCOL; see Box 6.1) and biogas systems in off-grid areas. It supported implementation by encouraging banks to channel their lending through microfinance providers with good rural links while also developing their own branch networks, which would have the longer-term effect of further reducing the cost to the end user. Under the same scheme, commercial banks also provide finance for Solar Irrigation Pumps directly to farmers' cooperatives, which are able to access favourable rates by combining their members' collateral.

The Bank's refinancing scheme provides loans at an interest rate of 5 per cent to commercial banks, who then lend to investors or households at a rate of 9 per cent. By 2010 the terms of the scheme required commercial banks to allocate 5 per cent of their loans to green lending, in return for low-interest credit with a short-term repayment period of four to five years.

The Development Bank of Ethiopia

The Development Bank of Ethiopia implements a range of programmes, including Market Development for Renewable Energy and Energy Efficiency Products. This is a World Bank-funded initiative aimed at encouraging investment in renewable energy technology and energy efficiency products.

The Bank uses a variety of instruments to encourage investment, including long-term loans and guarantees. It can lend developers 70 per cent of working capital at 8.5 per cent interest over a five-year period. It also provides concessional loans to microfinance institutions at 6 per cent interest over a ten-year period. This relatively long repayment period creates an incentive for providers to lend onward to low-income borrowers over shorter periods, and revolve the repaid funds as new loans. This credit helps private enterprises and households to invest in technologies such as biogas digesters, improved cooking stoves and solar lanterns.

The Development Bank of Rwanda (Banque Rwandaise de Développement)

Rwanda has set up a national basket fund called FONERWA (sometimes also called the Environment and Climate Change Fund, in translation). The Development Bank of Rwanda plays a key role in channelling LCRD finance from FONERWA to companies investing in renewable energy technologies, in the form of concessional loans: it receives loans from FONERWA at an interest rate of 2 per cent and lends onwards at a rate of 11.45 per cent. It also lends to private actors without this concessionary support from FONERWA, at a rate of 15 per cent. For comparison, market rates are around 18 or 19 per cent. (See Rai et al. 2015a; Rwirahira and Fisher 2015; Steinbach et al. 2015a for more details.)

Institutions such as these have been criticised for failing to reach the poorest. Often this is the result of a mandate that requires financial viability, which effectively limits them to using credit-based instruments rather than grants. However, some have now broken with this pattern in order to target the low-income communities (if not the extremely poor), incentivising commercial lending to borrowers usually considered too risky. The example of the Central Bank of Bangladesh, which was the first to take this kind of step (see Box 6.2), shows how a strong regulatory approach can be used to channel finance to marginalised communities via commercial banks and the private sector.

Discussion

Each of the two delivery models described has strengths and weaknesses. The integrated, 'one-stop shop' of special purpose agencies like IDCOL or the AEPC can create an enabling environment for reaching target communities, resulting in win-win opportunities for all actors in the value chain. They can support market creation, set up delivery networks, provide quality assurance, access to capital and training, and help new players to invest in low-income communities. Central and national development banks are not able to provide such support, but their regulatory authority can be used to construct a clear, phased mechanism to unlock finance that would not otherwise be available, engaging commercial actors more typically focused on mainstream markets. They are able to make this engagement mandatory, if necessary.

A key determinant of the effectiveness of any delivery model for LCRD, whether it is a special purpose agency or a national bank, is its ability to incentivise the use of intermediaries and financial instruments according to their relative usefulness in helping communities access finance. We look at this next.

Using combinations of intermediaries to reach low-income communities

All of the entities described – whether special purpose agencies or national banks – make use of multiple intermediaries in order to reach their target markets, which are often isolated in both geographic and economic terms. These intermediaries can be broadly categorised as 'financial agents' – such as commercial banks, microfinance institutions and NGOs – and 'delivery agents' – actors that deliver projects 'on the ground' such as manufacturers, suppliers, cooperatives, traders and NGOs (see Table 6.2).

Among the financial agents, commercial financial institutions are typically able to provide the scale of finance needed, but lack both the incentive to invest and the networks needed to reach target markets – the local knowledge, community presence and relationships. As discussed, national banks are aiming to unlock finance for LCRD from commercial sources through

Table 6.2 Intermediaries engaged to deliver LCRD and their characteristics

Intermediaries		Characteristics	
		+	−
Financial agents	Commercial financial institutions	Can provide large-scale finance	Lack of reach in low-income rural areas
	Microfinance institutions, NGOs	Reach and experience in low-income rural areas	High fees can increase interest rates charged to end users
Delivery agents	Private companies, suppliers	Cost-competent products Good after-sales service	
	Cooperatives, community groups	Unlike individuals, can provide risk guarantees and joint collateral; more investable Able to manage operation and maintenance of technologies Are able to get cheaper credit through 'direct' loans from banks	May lack cohesion May lack continuity and longevity

concessional loans and regulatory action. The Central Bank of Bangladesh, for instance, has developed a 'green banking' circular that obliges commercial banks to allocate 5 per cent of their lending to green investments, including renewable energy. The Bank's refinancing scheme also allows commercial banks to access low-interest capital for investment in green energy and effluent treatment plants, so increasing the profitability of this lending (Rai et al. 2015a).

In contrast, microfinance institutions and NGOs are often sought as intermediaries specifically because of their established reach in low-income and rural communities: their local market knowledge, their relationships with relevant agencies, groups and individuals, and their understanding of the barriers and risks specific to these markets (Steele et al. 2015b). IDCOL, for example, prefers working with these partner organisations because their offices located in rural areas and experience of running microcredit schemes mean that they are well equipped to manage credit disbursement and collection, and after-sales services. It has used microfinance institutions to assess household energy needs, and to install and service Solar Home Systems (Rai et al. 2015a). Commercial banks funded by the Central Bank of Bangladesh also prefer channelling finance through a credit linkage with microfinance institutions and NGOs. In Ethiopia, microfinance institutions have similarly expanded their role in the national development bank's renewable energy programme, and are now working directly with suppliers.

However, while they are effective in improving access to finance, microfinance institutions do not always provide the cheapest capital to low-income users. In fact there is little evidence that they are effective in lifting the poor out of poverty (Duvendack et al. 2011; Steele et al. 2015b). Their transaction costs tend to be high, which can increase interest rates for end users. The Central Bank of Bangladesh is currently trying to regulate the interest microfinance providers charge. Another option may be to use microfinance providers in the early stages of market development, then channel funds through more cost-effective intermediaries. As mainstream financiers become more experienced in working with rural markets, they may be able to provide low-interest loans directly to end users.

Among delivery agents in Nepal and Bangladesh – those involved in providing equipment and operational support for projects – suppliers are often seen as effective intermediaries, providing cost competence and high-quality after-sales services. In Bangladesh, some suppliers are now entering into agreements with financial institutions to provide credit and related services for the Solar Home System and Solar Irrigation Pump schemes, similarly to microfinance providers. That is, rather than simply supplying equipment to microfinance institutions and NGOs, companies have taken over the role of these intermediaries and are providing end users with credit.

Some financial institutions also prefer to lend to cooperatives and community groups, which, unlike individuals, can provide the necessary risk guarantees or joint collateral. Financing end users directly in this way also reduces the amount of interest they pay. This is the case for farmers' cooperatives in Bangladesh, who are able to obtain loans from commercial banks (funded in turn by the Central Bank of Bangladesh) for Solar Irrigation Pumps. The banks may provide these loans either indirectly through microfinance institutions and NGOs, or directly. Typically, the process involves a tripartite meeting between a farmers' cooperative, a microfinance institution or an NGO and a commercial bank, followed by the submission of a formal Solar Irrigation Pump proposal to the bank (made either directly by the cooperative or on its behalf, by the microfinance institution or NGO). Purchasing a Solar Irrigation Pump is often too expensive for individual farmers; a cooperative can provide joint collateral. Where finance can be provided directly to cooperatives by the bank instead of through a microfinance institution or NGO this further reduces the cost to the end user.

In summary, while the conditions and contexts for LCRD differ across the four countries, funding agencies' experiences of working with intermediaries illustrate some general themes relating to the political economy of climate finance: while microfinance providers may be better positioned to access target markets in low-income and rural populations, intermediaries such as commercial banks and grant-channelling NGOs are cheaper for end users. Given the right incentives, a combination of these intermediaries can deliver climate and development finance to the poorest parts of society. The best 'mix' will depend on the extent of market development, as well as the particular financial needs of the communities concerned.

Using a menu of financial instruments to reach low-income communities

In addition to an effective delivery model and suitable intermediaries, enabling low-income and off-grid populations to access affordable low carbon energy also depends on the availability of appropriate financial instruments. The kinds of high-interest loans traditionally used by financial institutions operating in mainstream markets tend to exclude lower-income groups: their cost makes them unaffordable to potential borrowers, while relatively lower profits, higher transaction costs and the limited credit track record of the poor make them unattractive to potential lenders. In this way, traditional market-oriented instruments have contributed to the failure of finance to reach those at the bottom of the income pyramid.

However, alternative instruments can help low-income groups to access affordable finance. These include grants, concessional loans, composite lending schemes and risk mitigation instruments. It is also possible to align incentives to harness a combination of instruments.

A range of such instruments has been deployed across the four case study countries, and used individually or in combination as needed. Grants are especially important in the early stages of transitioning to low carbon development, to stimulate pro-poor markets and provide finance for necessary activities that do not generate profit (Rai et al. 2015a; Steele et al. 2015a). In Bangladesh and Nepal, IDCOL and the AEPC, respectively, are using a range of grant-based instruments to promote energy access. These include 'pure' grants in the form of subsidies, technical assistance grants providing expertise and grants combined with loans to make them low interest. IDCOL, for example, provides a grant to households to cover the upfront cost of buying a Solar Home System, as well as technical assistance grants to develop the capacity of the partner organisations that install these systems. These grants are paid on delivery of certain outputs – for example system installation – which effectively ensures that purchasing power lies with households and performance risks with the private investors.

In Nepal, 40 per cent of the AEPC NRREP's US $170 million budget is being disbursed in the form of grants. These cover between 30 and 50 per cent of the cost of buying and installing technology, with the rest coming from concessional loans. In both countries grant funding has also been helpful in the early stages of projects, covering initial tasks like feasibility research, product development and technical assistance for capacity building, as well as helping to introduce new market players.

Grants are more appropriate in some situations than in others. They work well for projects that are not yet generating revenue, but they can also increase government expenditure or cause market distortion if investment is on the rise (Chiang et al. 2007; Steele et al. 2015a). Both IDCOL and the

AEPC have attempted to phase out subsidies for all but the extreme poor as markets have developed and project partners have achieved adequate capacity. Reductions in subsidy must come at the right time, however, and in the right places: even when a market for low carbon energy has become established, the very poor will often continue to need support. It is important to continue subsidies for the most marginalised groups, to cover their upfront costs and ensure that the anticipated long-term outcomes are realised.

Loans can be moderated by a grant element to bring them within reach of low-income groups (Steele et al. 2015a). They can also be provided on more favourable terms, including extended repayment periods, instalments tailored to income levels and a lower bar for qualification (especially with regard to collateral). Credit risk guarantees provided by national development banks may also encourage investors to channel capital to risky projects.

In the four case study countries, both national banks and special purpose agencies are providing concessional or low-interest loans to financial intermediaries (see Boxes 6.1 and 6.2). However, because of their particular institutional mandate the tendency is for national development banks to offer only loans, in isolation. In contrast, donor-supported entities such as IDCOL and the AEPC are able to combine loans with grants and technical services. The latter has proved more effective in reaching poorer communities, since grants can be used in the early stages to build required institutional capacities and support the various complementary activities needed to develop immature markets.

In addition to loans, tailored lending schemes have provided poorer communities with further opportunities to access finance. One such scheme in Bangladesh was originated by commercial banks. Otherwise reluctant to lend to poor farmers who want to invest in Solar Irrigation Pumps, banks have combined loans for pumps with loans for seeds and fertiliser, increasing their confidence that farm output will increase sufficiently for the farmers to repay their loans (Rai et al. 2015a).

In summary, it is clear from the experiences in Bangladesh, Ethiopia, Nepal and Rwanda that pro-poor financial instruments can enable low-income communities to obtain finance for renewable energy technologies. Successful approaches include combining grants with loans in new areas of investment, with grants then being substantially phased out as markets mature. Grants for the poorest sections of the community may be retained to ensure that these key target groups can access funding. More generally, innovative lending schemes can provide unbanked customers with access to finance.

Based on these examples, it seems that effective approaches to financing decentralised energy projects employ a combination of financial instruments, which are variously designed to meet the needs of the particular communities and income levels being targeted. In this context, the concern for policymakers becomes how to make the right set of options available, according to the stage of market development. This in turn means getting the right mix of incentives – the third of our arguments.

What types of incentives encourage investors to invest in low-income markets and households to purchase renewable energy technologies?

Recruiting investors to community projects that would not otherwise be financially attractive means providing incentives to the right people to make the right choices, at the right time. An actor's decisions are influenced by their incentives, including their mandate, overall function and responsibilities, as well as their knowledge base and resources, and the wider policy environment. The interplay of all these political economy factors determines how effective LCRD financing modalities are in reaching the vulnerable.

In our four case study countries a wide range of incentives have encouraged policymakers, practitioners, investors and communities to develop renewable energy projects (Table 6.3 provides a summary). Starting at the 'top' of the value chain, high-level policies, including government and fiscal targets, have prompted policymakers and practitioners to set up incentives for actors further downstream. In both Bangladesh and Ethiopia, energy access and

Table 6.3 Incentives for actors to invest in renewable energy

Actors	Incentives for investing in rural renewable energy
Special purpose agencies, central and national development banks	**Policy incentives**, e.g.: • Bangladesh – electricity for all by 2021, 10% of electricity from renewable sources by 2030 • Ethiopia – renewable energy ambitions in Growth and Transformation Plan, National Energy Plan and Climate Resilient Green Economy strategy
Commercial banks, other commercial financial institutions	• **Economic incentive**: concessional loans to commercial banks, microfinance institutes, suppliers and private companies for investments in pro-poor renewable energy projects • **Regulatory incentives**, e.g. green banking policy in Bangladesh requiring banks to lend for renewable energy projects
Microfinance institutions, NGOs, suppliers	• **Economic incentive**: concessional loans with long-term repayment, capital buy-down grants and institutional development grants
Suppliers	• **Economic incentives**, e.g. in Bangladesh, tax holidays, exemptions on imports and local manufacturing of renewable energy equipment
End users	**Socioeconomic incentives**: • Reduced-cost technologies • Buy-down grants to cover down payments • Availability of credit on affordable terms • Lower running costs and improved outcomes from using renewable energy technology, in comparison to conventional alternatives

related social development have been established as national policy goals, and provide the primary driver for investments in decentralised energy projects. Bangladesh's key policy goals include 'electricity for all' by 2021 and generating 10 per cent of electricity from renewable sources by 2030. Instead of backing grid extension in rural areas, in order to maximise the number of people reached the government decided to direct its subsidies to small-scale infrastructure. It was also eager to reduce reliance on imported diesel and natural gas, and agricultural subsidies, all of which pointed to renewable energy. Incentives included reduced import tariffs and lower taxes on renewable energy products.

Similarly, Ethiopia has developed policies that promote investment in renewable energy and energy efficiency, including in its GTP, National Energy Plan and CRGE strategy. The government has also introduced finance-enhancing regulatory measures such as feed-in tariffs and power purchase agreements to incentivise private sector investment in renewables. In this way, policy incentives have a trickle-down effect that can be harnessed to engage actors at all levels (Rai et al. 2015a).

Continuing down the value chain, such policy incentives have in turn set up economic incentives as drivers for potential investors including commercial banks, microfinance institutions and suppliers. In our case study countries, these economic incentives have taken the form of low-cost loans and institutional grants, fiscal incentives, tax holidays and reduced import tariffs. For instance, commercial banks in Nepal and Bangladesh have been incentivised to support renewable energy investments by the availability of concessional finance for on-lending to suppliers or microfinance institutions. Policy drivers also operate here, for example in the form of the requirement in Bangladesh for commercial financial institutions to allocate a proportion of lending to green investments.

At the next stage of the value chain, microfinance institutions, NGOs and suppliers in Bangladesh, Nepal and Ethiopia have been encouraged to engage in the renewable energy market by the provision of low-interest credit by development banks and special purpose agencies. Capacity incentives operate here, too, as these actors typically have long-standing experience in managing finance for low-income end users. Suppliers in Bangladesh have been given further economic incentives in the form of tax holidays and exemptions on imports, and local manufacturing of renewable energy technologies.

At the level of the end user, communities have identified clear socio-economic co-benefits of investing in low carbon energy. In Bangladesh, households and farmers have adopted Solar Home Systems and Solar Irrigation Pumps in recognition both of the cost savings that can be made, due to cheaper fuel and lower cultivation costs, and of the increased productivity that can be achieved. Similarly, in Rwanda and Ethiopia, solar lanterns and biogas digesters offer users opportunities to improve their health, education and income. Direct economic incentives also operate at this level, in the form of instruments such as buy-down grants and affordable loans.

It is important to note that rather than simply driving investment in low carbon development projects, these positive incentives promote development specifically in low-income communities. The experience and reach of microfinance institutions and NGOs represent capacity incentives, prompting national development banks and special purpose agencies to engage with them. Economic incentives have similarly prompted the choice of delivery agents: in the four case study countries private SMEs have typically been favoured over, say, microfinance institutions as suppliers because of factors such as their engineering skills, cost competence, quality of after-sales services and marketing capacities. They are keen to sell their products and can provide the necessary support in low-income renewable markets, which require robust products that need minimal maintenance, offered to users at competitive prices.

Again, national agencies have identified economic reasons to lend through commercial banks in preference to microfinance institutions once markets are sufficiently developed, in order to reduce costs to end users. They have also made strategic use of financial instruments, for example deploying grants to help to develop markets, then concessional loans to unlock private finance and ensure some level of commercial viability.

From these examples we see that incentives can be structured to prioritise the needs of the poor. This involves measures such as setting policy targets to shape the choices of actors lower down the value chain and using specific types of instruments to attract and support investment from suitable financial and delivery agents, and so ultimately to achieve sustained engagement among the target end users.

Ensuring finance is available to the poor and ultra-poor

As part of our research we interviewed stakeholders including investors, microfinance institutions and banks, and held focused group discussions with end users (for more details, see Steinbach et al. 2015b; Rai et al. 2015a; Rwirahira and Fisher 2015; Kaur et al. 2016). We then analysed the results to assess how different programmes and business models have been effective in

- targeting funds to reach the poor;
- using public funds to leverage finance for low-income populations from other sources;
- generating finance on appropriate terms, to meet the small-scale needs of those at the base of the income pyramid;
- facilitating co-benefits; translating LCRD into, for example, improved livelihoods, health and education.

The results are summarised in Table 6.4.

It is clear that some initiatives have been more successful than others in targeting low-income groups and generating finance that meets the specific

Table 6.4 Examples of extent of effectiveness of pro-poor measures from renewable energy projects in Bangladesh, Ethiopia, Nepal and Rwanda

Topic	Summary of findings from stakeholder interviews and discussions
Ways in which poor are being targeted	Most projects target rural, off-grid and remote areas, and are therefore considered pro-poor in principle. Upfront grants cover down payments which would otherwise be unaffordable; the remaining cost is then paid in instalments. Smaller, cheaper products are supplied to help to ensure affordability.
Ways in which finance is leveraged for low-income communities	Concessional loans help to unlock funds from commercial banks, microfinance institutions, suppliers and other private companies for investment in pro-poor renewable energy projects. Government tax incentives also motivate manufacturers and suppliers to enter the renewables market, using their own capital to start businesses. National banks are able to use their regulatory roles to encourage private sector investment.
Ways in which finance is delivered on appropriate terms	Providing intermediaries with flexible long-term repayment terms enables them to revolve funds and achieve better profit margins in relatively risky markets. Loan repayments are set according to the ability of buyers to pay. Monthly instalments kept below the equivalent cost of using kerosene, diesel or batteries or of payments for equivalent fossil fuel equipment.
Are the ultra-poor being left out?	Phasing out subsidies prevents market distortion but in cases where concessional loans are not combined with grants or subsidies the ultra-poor are unable to access finance. Funding models with strict requirements (e.g. collateral, single interest rate across income groups, unfavourable debt-to-equity ratio) tend to exclude the ultra-poor.

needs of the poor. Some models – mostly donor-funded special purpose agencies such as IDCOL and the AEPC – have specifically targeted low-income families by developing smaller-scale projects and by subsidising upfront costs to customers through a combination of grants and loans. IDCOL, for instance, has reduced the size of Solar Home Systems from an output of 130 watt-peak to 10–30 watt-peak to reduce their cost for low-income customers who are unable to afford bigger expensive products.

Central and national development banks, on the other hand, have simply offered low interest rates to renewable energy financiers such as commercial banks and microfinance institutions, with the implied assumption that this provision will trickle down to low-income markets. The reason for this difference in approach, as mentioned previously, is the differing institutional structures involved. A special purpose agency is able to pool resources and to experiment with the instruments, products and services it offers. The remit of a central or national development bank, however, places significant constraints on the approaches and instruments it can use. These banks are limited to predominantly loan-based instruments and are not in a position to provide the kinds of additional services that special purpose agencies can.

In terms of leveraging finance, the provision of low-interest credit through special purpose agencies and central and national development banks has incentivised private companies, microfinance institutions, suppliers and commercial banks to invest in pro-poor renewable energy projects. Flexible long-term repayment periods allow these intermediaries to revolve funds and achieve better profit margins; again, though, this is more evident with schemes offered by the special purpose agencies than those of national development banks. For example, IDCOL offers its intermediaries five- to seven-year loans, who in turn provide households with three-year loans. This allows the intermediaries to earn returns by revolving their loans twice. In contrast, the Central Bank of Bangladesh provides low-interest loans to financial institutions with a repayment period of just four years, which gives them only enough time to revolve the loan once.

Government policies such as tax incentives and low-cost finance have also been used to motivate manufacturers and suppliers to enter the renewables market, using their own capital to start businesses. In Bangladesh, for example, SMEs investing in renewable energy have been given tax holidays. National banks are also able to use their regulatory roles to encourage private sector investment.

A further issue in targeting the extreme poor is whether the finance provided is appropriate to their needs. Although all the institutions in our case studies – both special purpose agencies and national banks – have recruited various actors to low-income rural markets, the ultra-poor are still not adequately served (Rai et al. 2015a). For example, the Central Bank of Bangladesh's Solar Home System and Solar Irrigation Pump schemes require end users to cover upfront costs, provide collateral and repay their loans; repayment periods and interest rates are the same for everyone, and in the case of Solar Irrigation Pumps the debt-to-equity ratio depends on the relationship between bank and customer. While a proportion of the low-income population may be able to fulfil these requirements, the ultra-poor cannot. At present, there are no specific measures in place to provide for this group.

Efforts have been made to improve affordability: the monthly instalment for a Solar Home System is deliberately kept below the equivalent cost of using kerosene, diesel or batteries, and the Central Bank of Bangladesh also encourages solar panel suppliers to reduce the system price through its refinancing facility. Similarly, the payment terms for Solar Irrigation Pumps are purposely made more attractive than those for diesel pumps. Along with lower overall costs and payment schedules that are both more predictable and a better fit with farmers' income cycles, this means that farmers generally prefer the solar pumps.

In this case the bank's use of loans is effective in terms of financial viability. However, newer markets often require grant funding to support their initial growth. To achieve sustainable LCRD, the bank and participating financial institutions need to provide resources for establishing and monitoring quality standards for equipment, training financial institutions and covering end users' upfront costs, for example (Rai et al. 2015a).

A similar failure to serve the extreme poor may be unfolding in IDCOL and AEPC programmes in Bangladesh and Nepal, respectively. The price of renewable energy technology has fallen significantly and systems are increasingly affordable for low-income customers. The logic of the agencies' phased approach would suggest that subsidies can begin to be reduced in order to avoid market distortion. However, the ultra-poor still find it difficult to make down payments and monthly instalments. Withdrawing support at this point may compromise the programmes' effectiveness.

Particularly in Nepal, although the AEPC's subsidy model has improved energy access, progress has been slower than expected and the credit-based financial instruments available under the new Central Renewable Energy Fund may not be appropriate for the poorest. Many of the households interviewed during our research indicated that even now they have difficulty in accessing subsidies. In addition, these subsidies usually cover only 30–50 per cent of the cost and many households did not have access to sufficient credit or personal capital to cover the remaining amount (Steinbach et al. 2015b).

Our analysis indicates that the different funding entities' remits and sources of funding influence their ability to leverage finance, make it available on appropriate terms and adequately target the poorest of the poor. Special purpose agencies – funded by multiple donors and public funds and able to use a combination of instruments – are better able to develop approaches that meet the needs of the poor. In contrast central and national development banks are required to ensure basic commercial viability and therefore are largely limited to using loan-based instruments. As a result they have a difficult time tailoring products to the needs of the poor and ultra-poor.

Conclusions

In this chapter we have reflected on how energy access among the poor and ultra-poor can be improved through use of effective financing systems, intermediaries and instruments, and by providing suitable incentives. As the examples from our four case study countries show, a key element of effective pro-poor finance is the availability of appropriate delivery channels; actors who are well placed to mobilise finance for the poorest may not be equally well placed to deliver it. We have also been able to gain a general picture of what effective financing mechanisms for LCRD look like. They are likely to involve a range of intermediaries and financial instruments, deployed so as to achieve specific, cost-effective targeting of poorer segments of the population by structuring incentives to prioritise their needs.

In relation to the intermediaries involved:

- First, institutions with a strong regulatory command-and-control approach, such as central and national development banks, can be effective in engaging commercial banks and the private sector to channel finance to otherwise marginalised low-income and rural communities.
- Second, special purpose agencies such as IDCOL in Bangladesh and the AEPC in Nepal can use their greater institutional flexibility to choose methods of generating and channelling finance that match the specific needs of low-income consumers. For example, they are able to blend grants with loan-based finance and to provide additional technical assistance.
- Third, microfinance institutions and NGOs are well placed to work with low-income communities in the earlier stages of market development. In addition to their reach, their social development orientation incentivises them to make higher-risk investments in poorer communities, where commercial lenders' profit orientation acts as a disincentive. However, transaction fees can be high and mechanisms are needed to ensure the finance they offer is affordable. One option may be to phase out their role once other financial institutions have established a presence in the market; another would be for national banks to regulate the interest charged.
- This leads to our fourth point, which is that commercial banks may be better positioned to provide low-interest loans as markets mature (as seen in Bangladesh). As they set up more branches and their reach improves, they can channel cheaper capital directly to end users. They can also provide larger-scale finance, although again their profit motive may limit their willingness to invest in riskier markets.
- Finally, where product costs are high, small-scale users can gain access through cooperatives, as is the case with farmers purchasing Solar Irrigation Pumps in Bangladesh. Cooperatives are able to give the necessary risk guarantees and obtain direct financing from commercial lenders (instead of indirect financing through microfinance institutions), so reducing the cost of borrowing.

In relation to the financial instruments used:

- First, grants are crucial in the initial stages for supporting non-revenue-generating activities that build capacity in novel low-income renewable markets, including feasibility research, product development and technical assistance. They can also be used to improve reach through subsidising high-interest loans and reducing upfront costs for end users, or creating market incentives for SMEs. Where grants are used for general market development like this, they should be phased out over time to avoid market distortion (although their use should continue in order to support access for the poorest; see our fourth point).

- Second, grants need to be blended with concessional loans to provide scaled-up and long-term finance for end users with limited access to affordable mainstream finance.
- Third, the terms on which finance is offered to lower-income customers should be appropriate to their particular needs, which are typically for long-term, flexible loans with affordable repayment terms and limited security requirements.
- Fourth, even on such terms some of the poorest sections of society are likely to remain excluded. Even microfinance providers, which in principle target poor communities, tend to avoid offering finance to the ultra-poor, due to their inability to cover upfront costs and poor track record of repayment. For the most marginalised groups, therefore, targeted social protection instruments and safety-net programmes may be more appropriate than microcredit or concessional loans. Innovations in lending may also have a part to play in improving communities' purchasing power, following on from the example of commercial banks in Bangladesh offering 'composite' lending in the form of combined crop and solar irrigation pump loans.
- Finally, renewable energy systems and products can be tailored in ways that give poorer people greater access to them. For example, it may be possible to reduce the size of a system and thereby its cost, while still meeting household energy needs.

Achieving the right mix of intermediaries and instruments depends in turn on having the right incentives in place:

- Higher-level policy incentives are key in establishing further incentives all along the value chain. Enabling decentralised renewable energy access requires strong political will, and policies, targets and fiscal measures that communicate to actors at all levels, so incentivising the engagement of financial intermediaries, microfinance institutions and SMEs, and end users.
- Regulatory incentives can be used to recruit the private sector – commercial financial institutions, for example – that would otherwise regard investment in rural renewables as too risky.
- Similarly, economic incentives based on concessional financing are crucial in encouraging financial intermediaries such as private companies, commercial banks and microfinance institutions to invest. Suppliers also require economic incentives; measures such as tax holidays and reduced import duties can help signal market stability.
- End users need incentives to sustain their use of renewables. The most important is having access to finance on appropriate terms.

These conclusions based on understanding political economy dimensions of climate finance – particularly actors and their incentives – can be used as a starting point for formulating principles and strategies for LCRD, as a step towards ensuring projects achieve their purpose in benefitting the poorest and those most vulnerable to climate change.

References

Asaduzzaman, M., Yunus, M., Enamul Haque, A. K., Abdul Malek Azad, A. K. M., Neelormi, S. and Hossain, Md. A. (2013) *Power from the sun: an evaluation of institutional effectiveness and impact of Solar Home Systems in Bangladesh (A report submitted to the World Bank, Washington, DC)*. Dhaka: Bangladesh Institute of Development Studies. Available at http://sun-connect-news.org/fileadmin/DATEIEN/Dateien/New/Bangladesh_Idcol_Assessment.pdf (accessed 15 March 2016).

Ashden and Christian Aid (2014) *Lessons on supporting energy access enterprises: the most effective ways donors, investors, technical assistance providers and policy makers can support enterprises that provide access to clean energy*. London: Ashden and Christian Aid. Available at https://www.ashden.org/files/pdfs/reports/Ashden-ChristianAid_Report.pdf (accessed 19 January 2016).

Central Statistical Agency (2012) *Ethiopia: demographic and health survey 2011*. Addis Ababa: Central Statistical Agency. Available at https://dhsprogram.com/pubs/pdf/FR255/FR255.pdf (accessed 19 January 2016).

Chiang, E., Hauge, J. and Jamison, M. (2007) *Subsidies and distorted markets: do telecom subsidies affect competition?* Working Paper No. 7002. Boca Raton, FL: Department of Economics, College of Business, Florida Atlantic University.

Christensen, K., Raihan, S., Ahsan, R., Nasir Uddin, A. M., Ahmed, C. S. and Wright, H. (2012) *Financing local adaptation: ensuring access for the climate vulnerable in Bangladesh*. Dhaka: ActionAid Bangladesh. Available at http://www.actionaid.org/sites/files/actionaid/financing_local_adaptation.pdf (accessed 19 January 2016).

Climate Funds Update (2015) Climate Funds Update database. Available at http://www.climatefundsupdate.org/data (acessed 19 January 2016).

Duvendack, M., Palmer-Jones, R., Copestake, J. G., Hooper, L., Loke, Y. and Rao, N. (2011) *What is the evidence of the impact of microfinance on the well-being of poor people?* London: EPPI-Centre, Social Science Research Unit, Institute of Education, University of London. Available at http://r4d.dfid.gov.uk/PDF/Outputs/SystematicReviews/Microfinance2011Duvendackreport.pdf (accessed 21 January 2016).

Glemarec, Y. (2012) Financing off-grid sustainable energy access for the poor. *Energy Policy*, 47(1): 87–93.

Government of Rwanda (2013) *Economic Development and Poverty Reduction Strategy II*. Kigali: Government of Rwanda.

International Energy Agency (2012) *World energy outlook 2012*. Paris: International Energy Agency. Available at http://www.iea.org/publications/freepublications/publication/world-energy-outlook-2012.html (accessed 15 March 2016).

Islam, S. M. F. (Deputy CEO of IDCOL) (2014) *Financing renewable energy in Bangladesh*. Presentation at ESI Africa. Available at http://www.esi-africa.com/wp-content/uploads/2014/11/Formanul-Islam-Bangladesh.pdf.

Kaur, N., Rwirahira, J., Fikreyesus, D., Rai, N. and Fisher, S. (2014) *Financing a transition to climate-resilient green economies*. IIED Briefing. London: IIED. Available at http://pubs.iied.org/pdfs/17228IIED.pdf (accessed 26 March 2016).

Kaur, N., Tesfaye, L. Mamuye, S. and Fikreyesus, D. (2016) *Financing inclusive low-carbon resilient development: the role of national financial institutions in Ethiopia's Climate Resilient Green Economy*. IIED Country Report. London: IIED. Available at http://pubs.iied.org/10138IIED (accessed 25 March 2016).

Prahalad, C. K. and Hart, L. S. (2004). *The fortune at the bottom of the pyramid: eradicating poverty through profits*. Philadelphia, PA: Wharton School Publishing.

Pueyo, A., Gonzalez, F., Dent, C. and DeMartino, S. (2013) *The evidence of benefits for poor people of increased renewable electricity capacity: literature review*. Brighton: Institute of Development Studies. Available at http://opendocs.ids.ac.uk/opendocs/bitstream/handle/123456789/2961/ER31%20Final%20Online.pdf?sequence (accessed 19 January 2016).

Rai, N., Iqbal, A., Zareen, A., Mahmood, T., Muzammil, M., Huq, S. and Elahi, N. (2015a) *Financing inclusive low-carbon resilient development: role of Central Bank of Bangladesh and Infrastructure Development Company Limited*. IIED Country Report. London: IIED. Available at http://pubs.iied.org/10139IIED.html (accessed 19 January 2016).

Rai, N., Walters, T., Easterly, S., Cox, S., Muzammil, M., Mahmood, T., Kaur, N., Tesfaye, L., Mamuye, S., Knuckles, J., Morris, J., de Been, M., Steinbach, D., Acharya, S., Chhettri, R. and Bhushal, R. (2015b) *Policies to spur energy access: Vol. 2: Case studies of public-private models to finance decentralized electricity access*. Golden, CO: National Renewable Energy Laboratory.

Rwirahira, J. and Fisher, S. (2015) *Financing inclusive low-carbon resilient development: case studies of the Development Bank of Rwanda and the National Domestic Biogas Programme*. IIED Country Report. London: IIED. Available at http://pubs.iied.org/10150IIED.html (accessed 15 March 2016).

Sharma, A., Müller, B. and Roy, P. (2015) *Consolidation and devolution of national climate finance: the case of India*. Oxford: European Capacity Building Initiative. Available at: http://www.oxfordclimatepolicy.org/publications/documents/Consolidation_and_Devolution_final.pdf (accessed 15 March 2016).

Steele, P., Rai, N. and Nhantumbo, I. (2015a) *Beyond loans: instruments to ensure the poor access climate and development finance*. IIED Briefing Paper. London: IIED. Available at http://pubs.iied.org/17318IIED.html (accessed 21 January 2016).

Steele, P., Rai, N., Walnycki, A. and Nhantumbo, I. (2015b) *Delivering climate and development finance to the poorest: intermediaries that 'leave no-one behind'*. IIED Briefing Paper. London: IIED. Available at http://pubs.iied.org/17317IIED.html (accessed 19 January 2016).

Steinbach, D., Kaur, N. and Rai, N. (2015a) *Financing inclusive low-carbon resilient development in the least developed countries*. IIED Working Paper. London: IIED. Available at http://pubs.iied.org/10147IIED.html (accessed 15 March 2016).

Steinbach, D., Acharya, S., Bhusal, R., Chettri, R., Paudel, B. and Shrestha, K. (2015b) *Financing inclusive low-carbon resilient development: the role of the Alternative Energy Promotion Centre in Nepal*. IIED Country Report. London: IIED. Available at http://pubs.iied.org/10140IIED.html (accessed 15 March 2016).

UNDP (2011) *Towards an 'energy plus' approach for the poor: a review of good practices and lessons learned from Asia and the Pacific*. Bangkok: UNDP. Available at: http://www.undp.org/content/dam/undp/library/Environment%20and%20Energy/Sustainable%20Energy/EnergyPlusReport.pdf (accessed 19 January 2016).

Walters, T., Rai, N., Esterly, S. Cox, S. and Reber, T. (2015) *Policies to spur energy access*. Golden, CO: National Renewable Energy Laboratory.

Watson, J., Byrne, R., Jones, M., Tsang, F., Opazo, J., Fry, C. and Castle-Clarke, S. (2011) *What are the major barriers to increased use of modern energy services among the world's poorest people and are interventions to overcome these effective?* Collaboration for Environmental Evidence Review 11–004. Bangor: CEE. Available at http://r4d.dfid.gov.uk/pdf/outputs/systematicreviews/CEE11-004.pdf (accessed 19 January 2016).

Wilson, E., Rai, N. and Best, S. (2014) *Sharing the load: public and private sector roles in financing pro-poor energy access*. IIED Discussion Paper. London: IIED. Available at http://pubs.iied.org/16560IIED.html (accessed 19 January 2016).

World Economic Forum (2012) *Public-private roundtables at the third clean energy ministerial*. Cologny/Geneva: World Economic Forum. Available at http://www.cleanenergyministerial.org/Portals/2/pdfs/WEF_CEM3_Roundtables_2012.pdf (accessed 19 January 2016).

7 Using political economy analysis as a tool in national planning

Neha Rai and Erin J. Nash

Introduction

The empirical case studies in earlier chapters show how political economy factors – actors, knowledge, discourses and incentives – and the interactions between them can influence national decisions about the design, development and implementation of LCRD. Interactions are mediated by actors' incentives and the coalitions that they form to pursue their goals, as well as their perceptions of the challenges and opportunities at hand, which in turn influence which policy goals they pursue.

Unpacking these domestic political economy dynamics enables national governments to understand which factors can constrain and which can promote the effectiveness of policy processes and practice. Deployed at appropriate stages of the policy cycle, these factors can be used to improve both the planning and the implementation of LCRD. Political economy analysis, then, can be used as a tool to generate the knowledge needed to improve the design of LCRD policy, and it can be put to work to better understand and manage the barriers to successful implementation.

Despite the potential of this constructive use of political economy analysis, for many years it has been more usual to employ it as part of a critical approach; typically donors have taken a problem-driven perspective to understanding factors that constrain implementation of their programmes within a country or sector. Analyses undertaken from this perspective tend to focus little on how countries can use understanding of their own political economy to make better decisions with better outcomes (Fritz et al. 2009).

In this chapter we briefly discuss the evolution of analytical frameworks that make use of a political economy perspective, before moving on to consider the potential for a broader national planning approach. While policy cannot be set by fully rational decision makers working always for the public good and while policymakers may have vested interests, we argue that it is possible to modify and build on existing practice to enable governments to deploy political economy analysis as a constructive, practical tool for shaping and evaluating LCRD policies and programmes.

An assessment of existing political economy analysis tools

A wide range of development partners and donors have attempted to apply a political economy approach to understanding policy drivers, constraints and opportunities within a country. This section provides an overview of some of their work and illustrates that while the emphasis may have changed, political economy approaches continue to be used more as tools for donors to study country systems and diagnose problems than as a means of enabling countries to develop and improve their policy practices.

DFID's Drivers of Change approach

One of the donor agency initiatives for studying political economy at the national scale is the Drivers of Change approach, developed by DFID. Its aim is to understand how policy and institutional reforms emerge within a country, and to identify the factors that cause these processes either to stall, or alternatively to be sustained and successful (DFID 2009). The approach focuses on the interactions between three features of the political economy, and how these can be influenced to create incentives:

- **Structures** These are longer-term, larger-scale contextual features such as a country's natural resources, demographics, economic and social structures, geopolitical position, technological progress and climatic changes. There is generally little scope for influencing variables in this category, as they are typically determined by external forces and subject to gradual change.
- **Institutions** These are established laws, practices or customs. They can be formal – for example constitutional rules and distribution of powers, laws and election processes – or informal – political, social and cultural norms and obligations. Often these two exist in tension with each other. In places where formal institutions lack strength informal institutions are more likely to explain how and why things happen (DFID 2009).
- **Agents** Agents are actors who are internal to the policy process and country, including political leaders and parties, civil servants, business associations, trade unions and civil society organisations, as well some actors considered external to the state (and perhaps the policy process itself), for example foreign governments, regional organisations, donors and multinational corporations.

The interactions between these three features are dynamic, with cyclical effects (as depicted in Figure 7.1) occurring over a variety of timescales.

However, the Drivers of Change approach (in common with other early approaches to political economy analysis) provided only 'surface' explanations

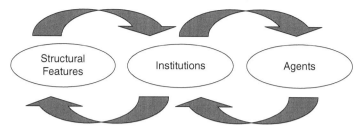

Figure 7.1 DFID's Drivers of Change approach
Source: Warrener (2004), available at http://www.gsdrc.org/docs/open/doc59.pdf, © DFID.

for how agents operate within the political system to bring about change (DFID 2009). DFID developed the Politics of Development framework to address this weakness.

DFID's Politics of Development framework

This method encourages a deeper and more systematic analysis of agents' relationships and the structural and institutional mechanisms they draw on to influence political decisions. The Politics of Development framework concentrates on four key elements within the decision-making process (DFID 2009):

- the broader historical, socioeconomic and cultural context, including the legitimacy of a process;
- the pressures from sections of the general population and interest groups, who influence political decisions indirectly;
- the formal and informal processes by which decisions are made;
- the ongoing politics associated with implementation, which can determine the success or failure of political decisions once made.

However, both of DFID's approaches – Drivers of Change and Politics of Development – were criticised as being strategies that, while they might help DFID advisers to gain a broad understanding of a particular national context, have limited usefulness for informing operational decisions. Subsequent approaches to political economy analysis focused much more on addressing specific issues and problems, becoming more practical tools (Fisher and Marquette 2014). Examples include the Swedish International Development Agency (Sida)'s Power Analysis tool (see Pettit 2013) and the World Bank's Governance and Political Economy approach. The former places strong emphasis on practical aspects of political economy, and involves assessing power relationships and their influence on different stages of development cooperation. The latter, meanwhile, is a problem-based diagnostic tool, used to identify the reasons for a particular programme's success or failure.

Sida's Power Analysis tool

Sida developed the Power Analysis tool as a means of mapping the informal political landscape and revealing the influence of its rules and structures on funding activities. It seeks to understand the power dynamics influencing different phases of development cooperation in a country. These phases might include, for example, scoping context, establishing goals and strategies, developing a project or programme, identifying partners and demonstrating outcomes (Pettit 2013).

Sida has used this approach in several countries with the aim of generating ideas for the development of cooperation strategies. It was used in Burkina Faso, for example, to analyse challenges and risks to cooperation related to democratisation processes, which included separation of powers, weakened opposition parties and the role of the country's army and traditional clan heads. In Tanzania, it was used to identify new entry points for building cooperation (Pettit 2013).

The World Bank's Governance and Political Economy approach

The World Bank has developed the Governance and Political Economy approach as a diagnostic tool for analysing specific problems and challenges. Applied at country, sector or project level, it seeks to (a) identify the problem; (b) map out relevant weaknesses in institutions and governance; and (c) understand the operation of deeper drivers within the political economy. The ultimate aim is to identify barriers to and opportunities for effective change (Fritz et al. 2009).

As an example, the World Bank used this approach to understand the issues within the mining sector in Mongolia, prior to implementing a technical assistance project. The analysis helped the country team to understand stakeholders' interests and incentives. It also revealed that a wide range of actors were effectively excluded from core stakeholder groups, and the consequent need for informed public dialogue regarding the sector's administration (Fritz et al. 2009).

The next generation of political economy analysis

As these examples show, the focus of political economy analysis has shifted from simply providing commentary towards providing practical solutions. However, these approaches have been developed by external donor agencies (who continue to be their main users) and inevitably they have generally taken on the perspective of these agencies. As a result analysis is typically oriented towards problem solving by external organisations, even the most recent analytical tools have been more critical than constructive. Fisher and Marquette characterise this form of political economy analysis as a 'risk diagnosis tool' that 'echoes the approach of the historic, technicalised "politics-free" approach to aid, keeping recipients at arm's length' (2014: 4).

Seeing the potential of an alternative approach, in this chapter we have adapted existing frameworks to provide a constructive tool for the use of governments and other stakeholders in the LDCs. This next-generation approach involves a more locally rooted analysis, one that can be used to develop and evaluate LCRD policies, and to build networks that are able to negotiate with donors and take plans forward to implementation.

Using political economy analysis as a constructive national planning tool

Political economy analysis can be used by a country's policymakers as a means of identifying the most effective pathways to successful LCRD. Depending on its purpose, analysis can be undertaken at different stages of the policy cycle, either retrospectively or prospectively. Retrospectively, a political economy analysis helps policymakers to identify why policies or interventions may not have gained traction or been successful in practice and what might be done differently in future. Prospectively, drilling down through the political economy layers can help policymakers to determine which actors and networks are best placed to deliver different aspects of LCRD. It can also be used to improve policy by providing a more in-depth understanding of what is likely to be effective in specific political and socio-economic contexts (Poole 2011). It can help policymakers to make appropriate decisions by providing them with a better grasp of stakeholders' particular goals, interests and incentives.

Therefore, rather than merely pointing to the barriers that different ideologies and incentives present to LCRD, political economy analysis can be a useful practical tool, outlining options for policies and actions more clearly (Fritz et al. 2009). A thorough knowledge of domestic political economy can provide policymakers with a clearer understanding of where collaboration, change and progress might emerge, allowing them to make more strategic choices.

It is important to note, however, that the political economy of LCRD involves a complicated network of causal relationships, running in different directions and with intricate feedback loops, making analysis a difficult (though in no way impossible) task. The structural and institutional features of a country's LCRD landscape will themselves be the products of previous policy decisions and interactions, and at the same time all structures, institutions and actors within this landscape will be evolving and changing continuously (Harris 2013).

Political economy analysis has given a range of planning tools a new currency. Depending on its purpose, a good analysis is likely to involve some combination of institution and governance mapping, stakeholder analysis, discourse analysis, network analysis, financial framework analysis and incentives analysis. Institutional and governance mapping involves identifying which actors have a role to play in the policy space. A more detailed stakeholder analysis is used to understand the particular roles and interests of these actors, as well as their level of influence. An example is provided in this book in the form of the case studies presented in Chapter 4, which use stakeholder analysis to examine the roles of decision makers in using CIFs at national level, in Bangladesh, Nepal and Ethiopia. Discourse analysis, meanwhile, is more retrospective, concerned with understanding why different actors' views converge on particular storylines, and how the networks that support particular decisions emerge. This approach is used in both Chapter 3 and Chapter 4 to analyse the role of discourse in national-level LCRD agenda setting.

In the remainder of this section we outline the typical steps in a political economy analysis; Figure 7.2 provides a summary. These steps represent a menu of options that can be tailored to the specific context and questions to be addressed; we illustrate this with examples taken from the case studies presented in Chapter 4. Note that some methods – such as institutional and governance mapping and stakeholder analysis – provide a rapid overview, while others – such as discourse and incentive analyses – are likely to form part of a more comprehensive study.

Step 1: identify the purpose and level of the analysis

The first question to ask is whether a political economy analysis is to be deployed retrospectively or prospectively. The former can be used to assess the effectiveness of a policy, programme or project, the latter to plan and implement a new initiative. The level of the analysis in turn depends on which type it is: a retrospective analysis is made at project or programme level in order to understand how processes and decisions have led to particular outcomes. A prospective analysis is made at sector or country level in order to identify actors, strategies and incentives, and to help to shape planning and implementation.

Step 2: outcome mapping

The second step is to define the outcomes that are likely to be impacted by actors and their interests. This involves listing the decisions made in the course of a particular project – including the institutional practices involved and the types of projects, actors, financial channels and instruments deployed – and the outcomes (either expected or actual, depending on whether the analysis is prospective or retrospective). This initial mapping can then be used to analyse how political economy factors influence the project decisions and outcomes. Where the analysis is prospective, it can also be used to improve understanding of how incentives may be structured to influence outcomes.

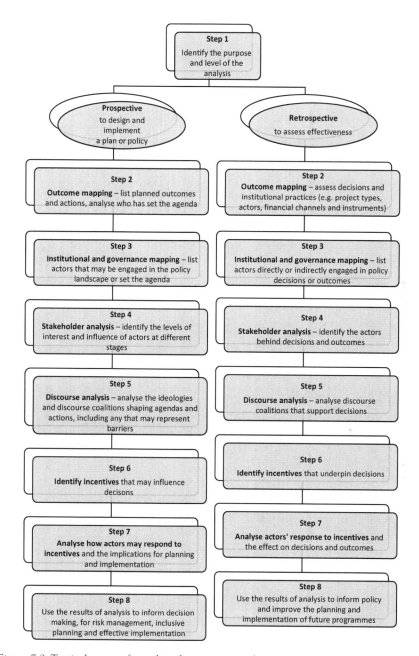

Figure 7.2 Typical steps of a political economy analysis

Step 3: institutional and governance mapping

The next step is to map the web of institutions and governance structures that influence how a policy or programme is developed and implemented. Three key benefits of this mapping are that it

- provides a more accurate and complete identification of the relevant organisations;
- helps policymakers to understand more fully the institutional and governance environments within which stakeholders operate, and how these settings shape stakeholders' interests and incentives;
- helps policymakers to understand where power rests within the policy process and to locate potential levers of change.

In this way, institutional and governance mapping can play an important role in helping governments gain a clear picture of capacities and arrangements, and of which institutions are best harnessed for the purposes of a particular intervention. For example, a government seeking to incentivise actors to engage with the Private Sector Window of the Green Climate Fund first needs to identify relevant private sector stakeholders and assess their capacities, as well as the institutional and administrative environment in which they operate.

It is important that governance structures are mapped alongside institutions, as it is these structures that determine how institutions and stakeholders interact. They consist of both formal rules and informal conventions; social and cultural practices as well as laws, regulations and administrative processes. It is also important to clarify how these formal and informal components interact: which carry more weight in a given situation, and the level of enforcement. Where there is a lack of enforcement, it is useful to identify the reasons for this, be they inadequate processes or a lack of awareness among stakeholders, for example. It is also useful to identify the mechanisms that exist to encourage integrity and accountability and limit corruption.

In a prospective analysis, institutional and governance mapping is used to obtain a clear picture of the actors likely to be involved in a policy or project and how they may influence decisions. In a retrospective analysis, it reveals the roles played by the different actors involved in decision making, both directly and indirectly. Policymakers and their advisors can then use this information to assess these roles in the light of outcomes, judging which actors contributed to the success or failure of a project, and how.

As an example, Table 7.1 summarises the results of a governance and institutional mapping exercise. It shows the institutions and governance arrangements in place for implementing the PPCR; one of the CIFs) in Bangladesh and Nepal (see Chapter 4 for more details). The analysis indicates that while Bangladesh has harnessed existing institutions to implement the PPCR, Nepal has developed new ones. Examining the interactions between governance arrangements, incentives and outcomes, we identified that the existing

Table 7.1 Institutions and governance arrangements for implementing the PPCR in Bangladesh and Nepal

Country	Institutional planning and coordination mechanism	Arrangements for implementation		Analysis
	Focal ministries	*Multilateral development banks*	*Counterpart line ministries/ departments*	
Bangladesh	Ministry of Environment and Forest (MOEF) Ministry of Finance **Institutional arrangement:** Harness existing mechanism set up for BCCRF within MOEF	World Bank Asian Development Bank International Finance Corporation	Water Development Board Local Government Engineering Department Previously Ministry of Agriculture Now MOEF	Bangladesh did not set up a dedicated agency for implementing the PPCR. As a result there was inadequate clarity on roles and responsibilities. Implementation arrangements between multilateral development banks and counterpart ministries (e.g. the World Bank and the Water Development Board) strengthened long-standing partnerships. This led to prioritisation of business-as-usual pipeline coastal embankment projects.

Table 7.1 continued

Country	Institutional planning and coordination mechanism	Arrangements for implementation			Analysis
	Focal ministries	Multilateral development banks	Counterpart line ministries/ departments		
Nepal	Ministry of Science, Technology and Environment (MOSTE) **Institutional arrangement:** New institutions within MOSTE: • Two coordination committees: PPCR coordination committee chaired by MOSTE and the National Planning Commission, and the Multi-stakeholder Climate Change Coordination Committee • Climate Change Programme Results Framework Coordination Committee – facilitates coordination between committees	Asian Development Bank	Department of Soil Conservation and Watershed Management		Nepal used technical assistance funds to set up new institutions to implement the PPCR.
		International Finance Corporation	MOSTE		
		World Bank	Department of Hydrology and Meteorology Ministry of Agriculture Development		

Source: Adapted from Rai et al. (2015a).

arrangements governing the interaction between multilateral development banks and line ministries had influenced the decision to invest in existing 'pipeline' projects (Rai et al. 2015a; see also Chapter 4).

Step 4: stakeholder analysis

Over the short to medium term, the key interactions likely to play out at national level are those between stakeholders and institutions, rules and norms. Stakeholders shape institutions – especially where those institutions are weak or volatile – while institutions influence the incentives and constraints that stakeholders are subject to (Fritz et al. 2009). A stakeholder analysis is a systematic qualitative exercise undertaken to identify relevant stakeholders and the roles they play, or could potentially play, at different stages of the policy cycle – in planning, institutionalisation, implementation and evaluation. Stakeholders are individuals or groups pursuing particular interests. They include political parties, government ministries, local government, the judiciary, the military, business associations and private sector companies, NGOs, religious organisations, trade unions, farmers' associations, the media, academia, external donors, foreign investors, foreign governments, and regional or international organisations.

The first step in a stakeholder analysis is to list all the actors involved in the policy process. Often the list is very long; however, it can be made more manageable by grouping stakeholders together. There are many ways of doing this, but common categorisations include demand- and supply-side stakeholders, champions and opponents of reform, stakeholders directly and indirectly involved in decision making, and 'winners', 'neutrals' and 'losers'. Second-order categorisations – for example grouping by function or sector – can provide further clarification. Care should be taken to use categories that are meaningful and unambiguous, so that the analysis reveals rather than obscures the underlying dynamics.

Table 7.2 provides an example of stakeholder categorisation. It lists stakeholders according to whether their involvement in the PPCR process in Bangladesh and Nepal is direct or indirect. It indicates that actors from civil society and the private sector were not directly involved, whereas core government ministries (finance and environment) and line ministries and multilateral development banks had prominent roles in the design and implementation of PPCR programmes.

Once all the stakeholders have been identified, the next step is to map them according to their levels of interest and influence at each stage of the policy process. Again, there are a variety of ways of doing this, some complex and others very simple. One common approach involves assigning stakeholders to a grid quadrant by plotting their level of interest on one axis and their level of influence on another (see Figure 7.3). The variables shown on the axes can be modified, as needed. Other common choices include the level of impact on the policy process, the level of support for LCRD, and attitude to change.

Table 7.2 Categories of stakeholders engaged in the design and implementation of the PPCR in Bangladesh and Nepal

Involvement	Categories	Institution types
Direct	Government core ministries	Ministry of Finance Environment Ministry
	Government executing line departments	Line departments directly implementing CIF
	Multilateral development banks	Actors directly involved in delivering CIF
Indirect	Other government actors	Government actors not directly involved in CIF delivery or those who have fallen out during the implementation stage
	Other multilateral actors	Organisations such as UNDP
	Civil society	Civil society organisations NGOs Think tanks
	Private sector	Private sector associations Relevant independent private companies
	Bilateral agencies	Donors and development partners

Source: Rai et al. (2015a), available at http://pubs.iied.org/pdfs/10111IIED.pdf, © IIED.

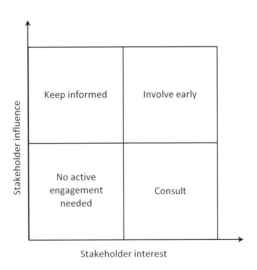

Figure 7.3 A power-interest grid for stakeholder mapping

Political economy analysis as a national tool 143

The position of each stakeholder on a power-influence grid indicates their relative importance within the policy process. Those stakeholders closer to the top right-hand corner have a combination of interest and influence and can cause a project to either succeed or fail. It is also important to note that actors' roles, interests and level of influence can change, although those with high levels of interest and influence are more likely to retain their position throughout the policy process (Bryson 1995).

The grid quadrants can be used to guide policymakers' interactions with stakeholders, helping to make their engagement more effective and strategic. For example, whereas high-interest, high-influence actors should be involved early on in the policy cycle (or otherwise have their expectations actively managed), those in the high-influence, low-interest quadrant might simply be kept informed.

Returning to our case study example, Figure 7.4 summarises a stakeholder analysis that maps actors involved in PPCR decisions in Bangladesh, based on data from semi-structured interviews. Analysis of these data involved

- categorising actors by their level of involvement in PPCR project selection, i.e. whether they were directly or indirectly involved;
- further categorising actors by sector or function, for example as multilateral development banks, donors, policy leaders or civil society organisations;
- evaluating the extent of actors' involvement and their levels of interest and influence at the various different stages of the policy cycle.

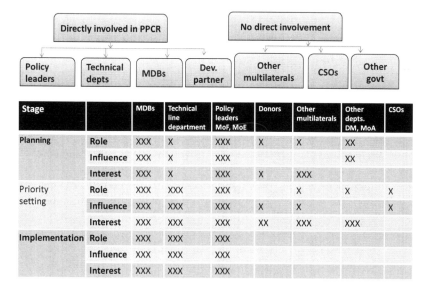

Stage		MDBs	Technical line department	Policy leaders MoF, MoE	Donors	Other multilaterals	Other depts. DM, MoA	CSOs
Planning	Role	XXX	X	XXX	X	X	XX	
	Influence	XXX	X	XXX			XX	
	Interest	XXX	X	XXX	X	XXX		
Priority setting	Role	XXX	XXX	XXX		X	X	X
	Influence	XXX	XXX	XXX	X	X		X
	Interest	XXX	XXX	XXX	XX	XXX	XXX	
Implementation	Role	XXX	XXX	XXX				
	Influence	XXX	XXX	XXX				
	Interest	XXX	XXX	XXX				

Figure 7.4 Results of a stakeholder analysis of the PPCR initiative in Bangladesh
Source: Based on data from Rai et al. (2015a).

The results show that some actors who were directly involved and had high levels of interest in PPCR planning and implementation were nevertheless excluded from decisions made during the priority setting and implementation stages, and that this was largely due to a lack of influence. Multilateral agencies, for example UN and civil society organisations, argued that community-based projects should be prioritised. However, their lack of influence was reflected in the type of projects finally selected, which were exclusively large infrastructure projects already in progress, managed by multilateral development banks and relevant government line ministries.

Step 5: using a discourse analysis approach to identify networks and coalitions that can support or hinder decision making

Stakeholders tend to become aligned to particular discourses; that is, to particular ways of thinking and talking about social and physical phenomena (Dryzek 2000). Coalitions may emerge around shared narratives; for example, in the context of LCRD a discourse coalition might coalesce around a common understanding of the core objectives of an initiative, and how it might best work in practice. These discourse coalitions, if concretised into policy networks, can influence the policy and investment decisions made, and their success (or otherwise) in practice. Even if they do not transition to become policy networks they can still influence decisions through the ways ideas and language are structured. For example, PPCR investment decisions in Bangladesh were driven by a network of actors that included multilateral development banks, core government ministries and influential line ministries that shared the view that 'transformational change' in national climate responses could be achieved by investing in large-scale infrastructure and economic growth. These actors were in a position to operationalise these views due to their central position within the decision-making process, as well as their resources and expertise.

When a narrative is widely shared but the actors supporting it are more marginal, it is less likely to translate into practice. A social development approach to achieving transformational change through PPCR investment in Bangladesh, which was supported by UN agencies and civil society bodies, is one example of this. The actors championing this viewpoint came from organisations that shared few points of contact or resources, and were in general less directly involved in PPCR decisions. There were few opportunities for coalition building and few shared incentives to take action, thus diluting their influence. On the other hand, it is possible for policymakers to act to support alternative views, deploying resources to strengthen consensus and facilitate synergies in order to move particular policies and programmes forward.

It is also important to manage actors who hold alternative positions that may hinder action. Where their views cannot be integrated into the consensus position, actors can become marginalised, disconnected from the wider stakeholder network, with the risk that they will set up barriers to implementation. An instance of this is provided by Bangladesh's PPCR-funded project to repair coastal embankments, which originally also included plans to improve surrounding areas of forest. The Forestry Department, however, felt excluded from initial consultations and this led to a lack of cooperation during the delivery stage. As a result the scope of the forestry plans was cut back. To avoid such hurdles, governments need to manage stakeholder interactions and expectations from an early point.

Where a discourse analysis is retrospective, it is used to identify which coalitions and narratives have helped to shape a particular decision or outcome, how and at what stage of the process, and the drivers underlying their emergence. In a prospective assessment, it identifies which coalitions and narratives are capable of either promoting or stalling a particular policy or action.

A range of analytical methods can be used to identify actor networks and their shared narratives. As discussed in Chapter 3, a common approach is to use Hajer's (1993) and Reissman's (2008) thematic discourse analysis frameworks to interpret data from semi-structured interviews with stakeholders. A typical framework has four main elements:

- **Policy discourse analysis or overarching narrative analysis** This involves analysing actors' narratives in terms of problem- and solution-framing, and their interpretation of policy objectives.
- **Identifying storylines** This is about looking for the 'snapshot' storylines that have become shorthand among actors for particular policy ideas.
- **Identifying shared narratives** This involves identifying similarities between different actors' accounts (Reissman 2008) and organising them into categories.
- **Identifying discourse coalitions** The aim here is to gain an understanding of where narratives are supported and by whom, within the actor network.

The final element in the process is identifying whether and when narratives become integrated into official policy.

As an example, discourse analysis undertaken for the Bangladesh study found that stakeholders involved in PPCR planning fell into three main groups, based on their views about whether and how PPCR projects could realise the objective of 'transformational change'. Some believed that the route to transformation was building climate resilient infrastructure, some looked to social innovation, and some were generally sceptical of the PPCR's transformational potential. These groupings are reflected in Figure 7.5.

The first of these coalitions – based on climate resilient infrastructure – consisted of core government ministries, executing line ministries and multilateral development banks. All were heavily involved in designing and delivering the PPCR. Meanwhile, as discussed, those stakeholders

Actor	Preferred narrative for PPCR's best route to 'transformational change'				
	Climate resilient infrastructure	Social innovation	Mainstreaming	Private sector inclusion	Sceptical
Government – core	●				
Government – line ministry	●				
Multilateral development bank	●				
Civil society					●
Other multilateral		●	●		
Government – other		●			●
Bilateral				●	●

Figure 7.5 Actors' narratives about how PPCR will bring about 'transformational change' in Bangladesh
Source: Rai et al. (2015a), available at http://pubs.iied.org/pdfs/10111IIED.pdf, © IIED.

advocating social innovation and inclusiveness did not manage to secure a firm role in the planning process. Similarly, scepticism about the PPCR – a view found throughout the stakeholder network – had little impact on the decisions made. Thus the coalition between government and the development banks drove the decision-making process.

Step 6: identify incentives

Incentives shape both stakeholders' narratives and their eventual decisions. They can also strengthen the position of some coalitions in the decision-making process and weaken that of others: a lack of incentives (or the presence of disincentives) helps to marginalise some coalitions, meaning that their narratives have less influence. An early diagnostic understanding of incentives in a prospective analysis can help policymakers to align policies and harness coalitions to achieve desirable outcomes. A retrospective analysis can give a holistic idea of actors' incentives and therefore the drivers that led to specific outcomes.

Data on the incentives and disincentives driving behaviour can be collected via semi-structured interviews with stakeholder representatives. Interview scripts can then be analysed and responses aggregated by categorising both stakeholders and the incentives they report. Example categorisations appear in Tables 7.2 and 7.3.

Actors can derive incentives from a range of sources, including their mandates, organisational structures, procedures, policies, resources and knowledge. Ultimately, though, incentives are generated by the institutional 'rules of the game'. That is, stakeholders are likely to behave differently in

Table 7.3 Summary of the main incentives driving PPCR priorities in Bangladesh

Country	Economic incentives	Policy incentives	Knowledge incentives
Bangladesh	<ul><li>Existing pipeline projects</li><li>Technology track record in infrastructure</li><li>Existing partnerships between multilateral development banks and line departments</li></ul>	<ul><li>Existing climate change policies identifying investment priorities (BCCSAP and NAPAs)</li></ul>	<ul><li>Vulnerability assessments</li><li>Loss and damage assessments</li></ul>

Source: Rai et al. (2015a), available at http://pubs.iied.org/pdfs/10111IIED.pdf, © IIED.

different institutional contexts, even though the interests they are pursuing are the same. This is because institutions, it is argued, cause some courses of action to appear more rational and appealing than others, in terms of helping actors to achieve the ends they value. Institutions generate and embody the incentive structures that actors must navigate. Shifts in institutional policy and practice generate new incentives and can bring about change. This is why development programmes often try to change actors' incentives rather than their interests (UNDP 2012). It is also why political economy analyses are focused on institutions.

Incentives can be categorised as:

- **Policy incentives** These are provided in the legislation, regulations or institutional mandates that support particular discourses and decisions. Note that incentives at policy level can in turn create further incentives lower down the value chain.
- **Economic incentives** These are resources, funds, technologies, and so on.
- **Knowledge and capacity incentives** This is the evidence and information, as well as the availability of technical skills, that generate understanding of a situation and compel one decision rather than another.
- **Reputation incentives** Actors or institutions may make decisions based on the idea that they can enhance their reputation and others' goodwill towards them, or alternatively that they will avoid reputational harm.
- **Socioeconomic incentives** A decision may be expected to result in positive socioeconomic changes such as improved livelihoods, better education and health or reduced inequality (Rai et al. 2015a; Steinbach et al. 2015).

Returning to our example case study, a diverse range of incentives contributed to the narratives, priorities and decisions relating to the PPCR in Bangladesh; these are summarised in Table 7.4. Resource and economic incentives strongly favoured investments in large-scale infrastructure. The multilateral development banks and government departments leading the planning and management of the programme had worked together on previous infrastructure initiatives, and this track record made the use of PPCR money for similar purposes very attractive. There was also co-finance available for top-up funding of coastal infrastructure projects already in the pipeline. The principal sources of policy incentives were Bangladesh's climate change strategy and its NAPA, both of which supported infrastructure-based initiatives. Knowledge incentives also played a key role: existing climate change vulnerability assessments, including loss and damage evaluations associated with Cyclone Sidr in 2007, recommended that US $1.2 billion should be invested in the rehabilitation of coastal embankments.

Step 7: analyse the interactions between actors' choices, discourses and incentives: the dimensions of political economy

Where political economy analysis is being used retrospectively to evaluate a specific LCRD policy or initiative, we need to understand the choices made, which actors and networks made and supported them, and how they have been embedded within institutions and policies. These features may be obvious, but it is also possible that a full picture will only emerge through investigation, for example via stakeholder interviews.

One of the many tools that can assist with this investigation is the climate finance landscape framework, as illustrated in Figure 7.6. This models climate finance as a chain of actors, and serves to locate and highlight the particular roles of the sources, users and uses of finance, and the financial intermediaries, instruments and planning systems involved in mobilising and distributing it.

An incentive analysis can then be used to understand actors' motivations and constraints, and how these led to particular actions. How and where LCRD approaches have been institutionalised should also be documented by undertaking an analysis of the relevant institutions and mechanisms of governance, both formal and informal.

How can political economy analysis inform policy and planning?

Any LCRD proposal can be expected to have its proponents and its opponents, with views and actions variously driven by ideologies, incentives and

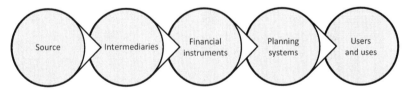

Figure 7.6 Climate finance landscape framework
Source: Adapted from Buchner et al. (2012); Kaur et al. (2014) and Rai et al. (2015b), available at http://pubs.iied.org/pdfs/10111IIED.pdf, © IIED.

resources that in turn have been shaped by the wider political economy. In order to take policy forward and avoid time-consuming disputes, governments and development partners need to navigate towards achieving consensus among a broad range of actors. Schmitz and Scoones, in their report *Accelerating Sustainability: Why Political Economy Matters* (2015), emphasise that this involves acknowledging those coalitions (which they term 'alliances') that seek to slow down decision making while fostering alignment and joint action among those that are supportive.

Understanding the political economy of LCRD will thus help national policymakers to drive decisions and implement them effectively, with less risk of stakeholders putting obstacles in the way. A political economy analysis can be used to map out the prevailing patterns of coalition and dissent, thus enabling them to respond appropriately.

Figure 7.7 shows a 'ladder' for applying the findings of a political economy analysis. First, resources can be channelled in the direction of those actors with a shared vision, thus strengthening coalitions and generating synergies that move the policy process forward.

Second, policy and economic incentives that encourage the integration of alternative and marginalised views should be actively pursued. In this way, a deeper understanding of the political economy context can help ensure that LCRD decisions take account of diverse views and are representative across stakeholder groups – and they may be made more effective and innovative, as a result. Another measure suggested by this inclusive approach is ensuring that actors with responsibility for delivering policy, such as line departments, are involved in the decisions that directly affect them. The aim here is to promote a sense of ownership and improve cooperation. The example given earlier – when a perceived lack of consultation led to poor cooperation between government departments involved in Bangladesh's coastal

Figure 7.7 General procedure for applying the findings of a political economy analysis
Source: Rai et al. (2015a), available at http://pubs.iied.org/pdfs/10111IIED.pdf, © IIED.

embankments project – illustrates the usefulness of more inclusive decision making in avoiding grievances.

Third, policymakers should seek out competing or conflicting views that have the potential to create obstacles down the line, in order to negotiate and manage stakeholders' expectations. This will help to reduce the risk of disruption, as well as providing a space in which objections and grievances can be discussed and responded to.

References

Bryson, J. (1995) *Strategic planning for public and nonprofit organisations*. Rev. edn. San Francisco, CA: Jossey-Bass.

Buchner, B., Falconer, A., Hervé-Mignucci, M. and Trabacchi, C. (2012) *The landscape of climate finance 2012*. Climate Policy Initiative report. San Francisco, CA: CPI.

DFID (2009) *Political economy analysis: how to note*. DFID Practice Paper. London: DFID. Available at http://www.odi.org/sites/odi.org.uk/files/odi-assets/events-documents/3797.pdf (accessed 22 December 2015).

Dryzek, J. S. (2000) *Deliberative democracy and beyond: liberals, critics, contestations*. Oxford: Oxford University Press.

Fisher, J. and Marquette, H. (2014) *Donors doing political economy analysis^{TM}: from process to product (and back again?)*. Developmental Leadership Program Research Paper 28. Birmingham: DLP. Available at http://publications.dlprog.org/Donors%20Doing%20Political%20Economy%20Analysis%20-%20From%20Process%20to%20Product%20(and%20Back%20Again).pdf (accessed 22 December 2015).

Fritz, V., Kaiser, K. and Levy, B. (2009) *Problem-driven governance and political economy analysis: good practice framework*. Washington, DC: World Bank.

Hajer, Maarten (1993) Discourse coalitions and the institutionalisation of practice: the case of acid rain in Great Britain. In F. Fischer and J. Forester (eds) *The Argumentative Turn in Policy Analysis and Planning*. Durham, NC and London: Duke University Press, pp. 43–67.

Harris, D. (2013) *Applied political economy analysis: a problem-driven framework*. London: ODI. Available at http://www.odi.org/sites/odi.org.uk/files/odi-assets/publications-opinion-files/8334.pdf (accessed 27 March 2016).

Kaur, N., Rwirahira, J., Fikreyesus, D., Rai, N. and Fisher, S. (2014) *Financing a transition to climate-resilient green economies*. IIED Briefing Paper. London: IIED. Available at http://pubs.iied.org/17228IIED.html (accessed 22 December 2015).

Pettit, J. (2013) *Power analysis: a practical guide*. Stockholm: Sida. Available at http://www.sida.se/contentassets/83f0232c5404440082c9762ba3107d55/power-analysis-a-practical-guide_3704.pdf (accessed 16 March 2016).

Poole, A. (2011) *Political economy assessments at sector and project levels*. World Bank How-To Note. Washington, DC: World Bank. Available at http://www.gsdrc.org/docs/open/PE1.pdf (accessed 27 March 2016).

Rai, N., Acharya, S., Bhushal, R., Chettri, R., Shamshudoha, Md. N. E., Kallore, M. E., Kaur, N., Neupane, S. and Tesfaye, L. (2015a) *Political economy of international climate finance: navigating decisions in PPCR and SREP*. IIED Working Paper. Available at http://pubs.iied.org/pdfs/10111IIED.pdf (accessed 22 December 2015).

Rai, N., Kaur, N., Greene, S., Wang, B. and Steele, P. (2015b) *Topic guide: a guide to national governance of climate finance*. London: Evidence on Demand. Available at http://www.evidenceondemand.info/topic-guide-a-guide-to-national-governance-of-climate-finance (accessed 22 December 2015).

Reissman, C.K. (2008) *Narrative methods for the human sciences*. Thousand Oaks, CA: Sage Publications.

Schmitz, H. and Scoones, I. (2015) *Accelerating sustainability: why political economy matters*. IDS Evidence Report No. 152. Brighton: Institute of Development Studies. Available at http://opendocs.ids.ac.uk/opendocs/bitstream/handle/123456789/7077/ER152_AcceleratingSustainabilityWhyPoliticalEconomyMatters.pdf;jsessionid=E58F6021E5D442FB5F87EBA9822611D2?sequence=1 (accessed 22 December 2015).

Steinbach, D., Acharya, S., Bhusal, R. P., Pandit Chettri, R., Paudel, B. and Shrestha, K. (2015) *Financing inclusive low-carbon resilient development: the role of the Alternative Energy Promotion Centre in Nepal*. IIED Country Report. London: IIED. Available at http://pubs.iied.org/10140IIED.html (accessed 22 December 2015).

UNDP (2012) *Institutional and context analysis: guidance note*. New York: UNDP. Available at http://www.undp.org/content/dam/undp/library/Democratic%20Governance/OGC/UNDP_Institutional%20and%20Context%20Analysis.pdf (accessed 27 March 2016).

Warrener, D. (2004) *The drivers of change approach*. ODI Synthesis Paper 3. London: ODI. Available at http://www.odi.org.uk/RAPID/Projects/R0219/docs/Synthesis_3_final.pdf (accessed 22 December 2015).

8 Supporting effective low carbon resilient development

Lessons from the political economy of the least developed countries

Susannah Fisher

Introduction

LCRD, a policy response to climate change, is a new and emerging agenda among the LDCs and poses particular public policy challenges. It is an important agenda in LDCs due to the high levels of poverty and climate change vulnerability in these states, making adaptation vital for continued development. Mitigation may offer potential synergies with adaptation and national priorities such as energy access and access to other sources of finance. LCRD offers a broad framing within which to bring together the adaptation, mitigation and development agendas. However, the reality of this agenda in the LDCs is relatively underexplored. This book has provided an empirically founded understanding of experiences in four LDCs and draws out lessons for other governments and actors interested in LCRD as an approach.

This requires more than a technical analysis, and more than looking at which policy is optimal or which scientific model should be used. It also requires an understanding of the decision-making process: what influences people to make particular decisions and how these decisions play out in reality, across sectors and at different scales. In this book we have used a political economy approach to address these issues, exploring the complex political and economic processes involved in national planning for LCRD, with the aim of understanding how ideas, power and resources interact to produce different outcomes (DFID 2009; Tanner and Allouche 2011).

In Chapter 1 we identified four main challenges or new dimensions to the LCRD agenda that have not been fully explored in the existing literature, and this book seeks to address these gaps. First, we saw that while the international dimensions of climate negotiations and interstate relationships have been studied by scholars across disciplines and have highlighted issues of equity in north-south relationships over climate change and international climate politics and justice, but there has been relatively little attention paid to the politics of the climate change agenda as it unfolds at national level in the global South (with some notable exceptions discussed in Chapter 2). Second, the LCRD agenda is new and poses particular institutional challenges, including issues of coordination and jurisdiction between and within scales and across the climate and development agendas; it also involves reconfiguring national relationships and structures. Third, the LDCs are among those countries that are most vulnerable to climate change; there is also significant inequality within their national contexts. Finally, analysis of national climate change policies and programmes has often been carried out from an external perspective, sometimes undertaken by development partners. This has meant that the findings have been less useful for supporting national implementation and decision making.

In order to better understand these four dimensions, this book focuses on:

- national experiences of climate change planning, rather than the international perspective that predominates in much of the literature;
- how countries are implementing their LCRD strategies and how this new agenda is playing out in practice;
- the LDCs, which are often left out of discussions;
- a constructive approach to research and analysis, one that can be used to support better LCRD planning and outcomes.

Chapters 3–6 are based on in-depth empirical research into four key dimensions of LCRD planning: national strategies; international climate funds; national climate finance; and the distribution and flows of finance to poorer communities. Chapter 7 looks at how to bring this knowledge into policy spaces. Each chapter also focuses on a particular dimension of political economy analysis.

- Chapter 3 looks at agenda-setting processes, analysing the role of discourses in realising potential synergies between adaptation, mitigation and development agendas, and the implications of this for future implementation.
- Chapter 4 examines a particular climate programme – the CIFs – in detail. It explores how actors, coalitions and incentives come together to support different approaches to energy security and climate change resilience, and considers the role of dominant and alternative narratives in supporting or hindering plans.
- Chapter 5 looks at emerging approaches to national climate finance and the potential for policy networks – stable groups of actors – to support the implementation of LCRD plans, as a means of addressing the need for cross-sector learning and coordination.
- Chapter 6 analyses how climate finance is channelled to low-income communities and households, and which instruments and intermediaries are best suited to this task. It also explores the incentives needed to reach the very poor.
- Chapter 7 discusses how to undertake a political economy analysis that can be used to support decision making.

In this chapter we use findings from across the book to draw conclusions about the political economy of climate change planning. Our conclusions relate to the three components of political economy identified in Chapter 1: actors and networks, discourses and knowledge, and resources and incentives. Finally, we draw on these conclusions to suggest lessons and ways forward for policy makers researchers and practitioners involved in national LCRD planning.

Actors and networks

One of the core challenges presented by the LCRD agenda is overlapping policy areas and jurisdictions in the national sphere: the emerging nature of the approach means that it does not fit within existing institutional relationships and silos. Understanding actors and the policy networks they work in will play a significant role in addressing this challenge. Another core challenge is the need to make decisions equitable and inclusive at all levels, from international to national to local. Who makes decisions, and with whom, is also an important factor.

Drawing on findings from across the book, we can make two main arguments relating to actors and networks. These relate to the engagement of appropriate actors and the harnessing of policy networks. We then highlight the implications of these findings for policy and practice, arguing that a range of actors needs to be included in LCRD processes to address issues of ownership and equity, and that the development of policy networks can support innovation in this new area of policy and promote learning about the best way forward.

Involving all appropriate actors

Involving all appropriate actors is key to the success of an LCRD plan, programme or intervention, as it addresses the particular challenge of policy coordination and inertia around initiating a new agenda. Which actors are appropriate depends on a number of factors including where convening power lies, who has 'ownership' of the agenda and where appropriate actors – those able to champion the agenda and drive it forward – are situated organisationally, as well as ensuring that the full range of diverse actors are engaged. Once these key actors are engaged, it is also important to provide time and space for building wider ownership.

Actors with convening power can play an important role in initiating the agenda and bringing diverse players together. Our research has identified a shift in climate change leadership within some governments, with responsibility being passed from ministries dealing with the environment to those dealing with (in particular) finance and planning. In Bangladesh, for example, responsibility for the country's involvement in the PPCR (one of the CIFs) is shared between the environment and finance ministries. The effect of this is to locate power and responsibility where there is greatest multi-sectoral authority. It also demonstrates wider 'buy-in' across government.

However, some governments have preferred to follow the more 'traditional' route, and place the responsibility with the environment ministry. This is the case in Nepal, which selected the MOSTE as its lead PPCR agency, despite a preference among some international actors for the MOFED (Rai 2013; Ayers et al. 2011). This move has helped to maintain a sense of national ownership. This is an important issue within LCRD planning: stakeholders in Bangladesh, Ethiopia and Rwanda all characterised having ownership and control over the LCRD agenda as a major incentive. Policy entrepreneurs or champions can also create support for a new agenda. High-level political figures were instrumental in initiating national strategies in Ethiopia and Rwanda, thereby increasing their political visibility.

As well as engaging actors with the appropriate convening power, i.e. those who feel a sense of ownership and those who can provide political leadership, it is also important to engage with the full breadth of actors – even those without convening power or authority – to ensure that considerations of equity and diverse perspectives are included right from the agenda-setting stage. We saw in Chapter 4 how minority interests not fully included within the agenda-setting phase of an initiative could impact on implementation, and the importance of bringing in alternative and more marginalised perspectives for equity and diversity of representation. For example, in Bangladesh, those not directly involved in the country's PPCR programme called for more attention to be paid to achieving social development, but this was not translated into PPCR decisions due to a range of economic and other incentives supporting other narratives.

The organisational context can influence how actors approach the issues at hand and the barriers they face to achieving LCRD objectives. This includes their ability to target the poorer sections of the community and thus address issues of equity and distribution of resources. We have seen that in the agenda-setting phase of LCRD there is often a range of views among government actors. This suggests that at this early point in policy development individual values and preferences tend to carry greater weight than organisational affiliation. However, as we progress through the policy phases towards implementation the influence of organisational position increases. For example, we found that in Bangladesh stakeholders from the government ministries responsible for the environment and energy split along organisational lines, supporting two different viewpoints.

In Chapter 5 we identified that actors have tended to favour a particular approach to mobilising and delivering climate finance, based on their policy mandates and capacity, i.e. in the organisational context. In Chapter 6 we discussed how climate finance could be used to address local energy needs. We saw that financial actors had particular strengths and weaknesses when it came to targeting poorer communities, depending on their organisational context: those with a strong regulatory command-and-control approach, such as central and national development banks, were effective in engaging commercial financial institutions and the private sector in this market. Special purpose agencies, on the other hand, were able to use their greater organisational flexibility to choose methods of generating and channelling finance in order to meet more fully the specific needs of low-income communities. Similarly, microfinance organisations and NGOs were generally better placed than commercial lenders to act as financial intermediaries, at least in the earlier stages of market development, due to their social development aims.

Involving appropriate actors throughout the policy process is an important way of building and maintaining a sense of ownership. In developing their national LCRD strategies, the governments of Ethiopia, Rwanda and Bangladesh all sought to establish cross-sectoral mechanisms to bring together government actors and to widen engagement within and beyond government. However, in many cases ownership remained with a small number of individuals or government bodies, and other stakeholders struggled to fully engage with the agenda. In Ethiopia and Rwanda stakeholders reported initially being unsure of what an LCRD agenda entailed and what their roles were. Similarly, lack of stakeholder engagement and buy-in with respect to Bangladesh's PPCR programme led to difficulties in implementation. In particular, the country's project to repair coastal embankments had originally included plans to improve the surrounding areas of forest. However, a lack of cooperation between the body responsible for implementing the repair project and the forestry department meant that these plans were scaled back. In Nepal an unanticipated imbalance of funding between the hydro-meteorology department and agriculture ministry led to the latter receiving less funding than anticipated, thereby creating conflict. These examples show the importance of engaging a range of stakeholders in developing and implementing new LCRD programmes.

In conclusion, while the makeup of the group of actors best placed to engage in a country's climate change planning depends on the context, our findings allow us to derive some broad principles to help to address the challenges of policy coordination and smooth the emergence of LCRD as a new agenda, as well as to support equity in decision making. Based on the findings discussed here, recruiting the appropriate actors – those able to drive the agenda forward – depends on where convening power is held, who has ownership of the agenda and where actors are organisationally situated. An actor's particular organisational position also influences how they engage with policies and their implementation; this needs to be taken into account along with their individual interests and preferences. Furthermore, engaging diverse stakeholders, including those without convening power or authority, is important to support LCRD policy processes and outcomes that are more equitable and inclusive. This needs to be supported by suitable incentives and resources, otherwise the views of more marginalised actors may not be fully heard. Finally, once the right actors are engaged, it is then important to provide time and space within the policy process for building wider ownership.

Harnessing policy networks

The emerging nature of the LCRD agenda means that it is not always supported by a policy community able to integrate it easily into existing tasks and relationships. New flows of climate finance are reconfiguring national policy landscapes, causing the balance of power between actors to shift. How aspects of the agenda might intersect – adaptation and mitigation, for example – is also a particular challenge. We argue that these challenges can be partly addressed by developing new policy networks – meaning relatively stable networks of actors built around common interests (Börzel 1998) – to support cooperation within and between sectors. This can help to provide the innovation needed at this new intersection of policies and sectors, as well as supporting learning and iterative improvement around the approach.

First, we show that these policy networks influence how an LCRD policy agenda emerges. Evidence from this book indicates that pre-existing policy networks are key in LCRD policymaking and implementation. For example, in Bangladesh PPCR funding was managed by those government departments already involved in climate-related programmes (see Chapter 4). In contrast, competing narratives and a lack of strong policy networks translated into a mix-and-match of investments in Nepal, involving energy technologies of different types. We have also seen that in the absence of a dominant consensus and strong policy networks there can be disagreements between actors and delays in implementation.

Second, we argue that it is possible to mitigate the impact of institutional policy silos and inertia by purposely supporting policy networks. Where discourse coalitions have emerged around policy approaches, these can be encouraged to develop into policy networks. Members of a discourse coalition do not necessarily see themselves as part of such a network; they may not work together and indeed may not even be aware of other actors with similar views. For example, the discourse coalitions forming around single- and multi-channel approaches to LCRD finance, as discussed in Chapter 5, have yet to evolve into policy networks. If they were to do so, this could promote information sharing and coordination between the two approaches, potentially enabling policymakers to leverage improved outcomes and put in place better solutions for resource mobilisation and delivery. This could also help to support dialogue and learning between national and international actors, enabling international financial intermediaries to deploy more appropriate incentives and instruments, and national actors to mobilise and deliver climate finance more effectively.

Discourses and knowledge

Actors' understanding of what the LCRD agenda means is still developing; they are still interpreting the concepts involved and the technical discussion that is ongoing in the international sphere. There is also the complex intersection between adaptation, mitigation and development to negotiate, and the challenge of making initiatives equitable and inclusive across national constituencies. We have used analysis of discourses and discourse coalitions to examine the circulation of ideas relating to LCRD, and in particular how and why the mitigation, adaptation and development agendas are being brought together within national contexts. This helps us to understand, for example, why particular technologies and renewable energy sources might be chosen over others. Such choices have social, political and environmental implications, in terms of who benefits from synergies between agendas (equity), who has the power to push agendas forward, and any potential maladaptive consequences. We make three arguments based on our findings, relating to the influence of national discourse on the LCRD agenda and priorities, of multiple and alternative storylines, and of marginalised storylines. We argue that identifying national discourses and priorities can help to clarify which dimensions of a policy or programme will find support, and also ensure that alternative or more marginalised storylines are not lost. A critical perspective is needed to ensure that issues of equity and inclusion are explicitly addressed in policy discourses.

One of the innovations of this book has been the use of storyline and discourse analysis of LCRD planning at national level to explore how ideas and opinions have shaped national policies. The term storylines refers here to the shorthand way in which actors understand and describe policy issues (see Chapters 1 and 3 for further details).

The influence of national discourses and priorities on the LCRD agenda

Chapter 3 showed how national discourses have shaped LCRD agenda setting in individual countries, creating a different emphasis in each. Bangladesh has prioritised adaptation, based on concerns about national ownership of the agenda, energy security and effective targeting of the poorest. In Rwanda the climate change agenda has become closely associated with a broader, nationally shared concept of environmental sustainability, while in Ethiopia it has become linked with the goal of transforming the national economy.

Similarly, in Chapter 4 we found that the dominant storylines within a particular programme were shaped by existing national priorities. For example, 'transformational change', as a core objective of the CIFs, had different national interpretations, and again these were consistent with national priorities. In Bangladesh transformational change was expected to result from infrastructure investments and economic growth, while in Nepal the focus was on long-term sustainability goals and the need for greater capacity. Storylines became dominant because they were backed by resources and incentives, and hence translated into investment decisions.

We can see, then, that national discourses and priorities shape agenda setting around national LCRD agendas or the dominant storylines within a particular programme. This suggests – despite international rhetoric to the contrary – that the economic and political impacts of adopting such an agenda may be incremental rather than transformative, as options that chime with existing policy discourses and incentives are the ones taken forward. Thus identifying national discourses and priorities concerning development can help to pinpoint which aspects of an LCRD agenda will gain traction with national stakeholders and may be reflected in policy decisions, and which dimensions might be more challenging.

The influence of multiple and alternative storylines

As discussed, policy coordination and representing the diversity of interests are challenging issues within LCRD planning. Understanding multiple storylines helps us to recognise how actors might coalesce around an approach (and therefore be better coordinated) and also how different interests and perspectives can be taken into account so as to support equitable implementation.

In Chapter 3 we found that national stakeholders support a range of different storylines, reflecting a range of different understandings of the LCRD agenda. In Bangladesh there was a particularly interesting divide between stakeholders from the government ministries responsible for the environment and energy, whose respective views coalesced around two distinct storylines, forming two distinct discourse coalitions. We also identified that this diversity of ideas extends beyond discourse, with multiple storylines becoming institutionalised – embedded in policy and financial decisions – in both Ethiopia and Bangladesh. Such diversity of views has important implications, to the extent that it may prevent coherent implementation; the risk is compounded as LCRD actions typically cut across traditional sectors, areas of policy and policy networks. Without a common understanding among stakeholders over what co-benefits might look like and how to realise synergies between adaptation and mitigation measures, there is a risk that trade-offs will not be addressed systematically. There could also be a risk of maladaptation in the long term.

Such differences in understanding can also lead to difficulties in implementation. Supporters of non-mainstream and dissenting storylines may stall progress or act as 'passive resisters'. As discussed in Chapter 4, one of the objectives of the PPCR, as implemented by the IFC, was to increase private sector investment in adaptation-related activities. In both Bangladesh and Nepal the intention was that this aspect of the programme would be implemented through the national ministries of agriculture (since it overlapped with their existing remit). However, officials within these ministries subscribed to an alternative narrative that saw the private sector as having insufficient capacity and did not support government funding of for-profit enterprises. In both countries this difference in framing has led to significant delays in implementation, and in both the IFC addressed this by bypassing the alternative narratives within the agriculture ministry and working with other actors. In Bangladesh the policy is now operated primarily by the environment ministry, while in Nepal it is directly managed by the IFC.

Diverse and minority views should be included from the early stages of policymaking. This ensures that such views are included before discourse coalitions have crystallised (and therefore before the formation of any policy networks), and strengthens stakeholder ownership of the process. This is also a way of opening up debate and uncovering ambiguities within policy. Identifying dominant and alternative discourses can help to identify where pitfalls might lie and where more time is needed to build up support (including incentives and resources) and consensus around a particular approach.

Storylines of trade-offs, equity and inclusion

Equity and inclusion are important dimensions of LCRD planning in LDCs. Analysing aspects of discourse that are marginalised can help us to identify where equity is not being fully considered, or where it is at risk of being sidelined in the move towards implementation.

We identified in Chapter 3 that some storylines may not feature prominently in stakeholder discourses, with the result that key questions relating to LCRD may not be addressed, either in debate or in the policies and practices that emerge. Policy trade-offs are an example of one such marginalised storyline; this important issue was largely marginalised from the agenda-setting phase of national planning discourses and policy frameworks in all of our focus countries. While in interviews stakeholders talked about policies that might imply trade-offs (for instance, energy choices, fertiliser use and modifying pastoralist livelihoods), they seldom conceptualised these as leading to potential trade-offs between, for example, growth and development. In Rwanda the overarching storyline of environmental sustainability does not fully tackle the issue of how options for mitigation or adaptation might be prioritised, and therefore what trade-offs might be necessary for different groups.

There are also questions concerning who benefits from particular storylines or approaches. The idea of a 'win-win' agenda is prevalent among many stakeholders, and does not necessarily leave space for considering who benefits and who is excluded. For example, who benefits from a low-carbon or green economy approach as opposed to an adaptation-first approach? This issue merits further examination.

As we discussed earlier, existing national priorities can shape agendas and priorities, which may already contain a commitment to progressive and equitable development outcomes. However, it is important to keep a critical eye on any new dimensions of equity that an LCRD agenda throws up, particularly those that could impact on vulnerable communities, such as trade-offs between long-term goals, short-term benefits and poverty alleviation.

Some of these issues of social equity and inclusion are highlighted by the experiences of SREP investments in Ethiopia and Nepal. In Ethiopia actors' views on the SREP's objective of transformational change coalesced around a dominant narrative – supported by government staff and multilateral banks – of economic growth, driven by improving energy security through diversification of on-grid technologies. An alternative narrative was articulated – by bilateral funders and the private sector – who argued that improving energy access for the rural poor would bring co-benefits for people and the environment, for example by helping to prevent deforestation, improve health and reduce indoor pollution. These actors were more marginal, however, and this view was not translated into investment decisions.

Narratives in Nepal addressed the same issue, with government policy-makers and multilateral development banks adhering to a storyline of transformational change through improved energy access. There were also narratives of economic growth and technology diversification – similar to the dominant narrative in Ethiopia – which were supported by a limited number of stakeholders from bilateral funders and multilateral development banks. Both these narratives were translated into investment decisions in Nepal, resulting in a range of investments, involving different types of on- and off-grid technologies.

The agenda-setting stage of the policy cycle is the time for big ideas, whereas the later stages require a more practical, grounded focus, being more concerned with decisions and programming. It is perhaps to be expected that issues such as trade-offs and 'winners and losers' have yet to be fully addressed in national LCRD planning processes, which are still in the early stages. However, any given policy approach will have different impacts on different sections of the population and ideas of climate justice stress the importance of an equitable approach within as well as between countries. It is important that debate over issues of equity and inclusion is initiated early on, to avoid making policy choices that are not fully inclusive and also to ensure that existing progressive national priorities are integrated with the newer dimensions of an LCRD approach.

Resources and incentives

Analysing incentives can help to address particular LCRD challenges concerning moving towards implementation (such as reconfiguring relationships, moving across scales, multi-sectorality, and long time frames) as well as helping to support an equitable and inclusive approach. Because LCRD is an emerging policy agenda, it has few existing incentives and resources to move it forward. Identifying the incentives that have helped to initiate and shape the approach in particular countries can help us to understand which parts of the approach are likely to be successful, which elements are likely to be dominant and which elements might need to be incentivised in other ways.

Our four main findings relating to this component of the political economy are concerned with incentives influencing finance, relationships and knowledge; with short- and long-term LCRD policy approaches; with aligning incentives across levels; and with structuring incentives so as to reach the poor.

Incentives relating to finance, relationships and knowledge

In Chapter 3 we saw that the main initial incentives for developing a national LCRD policy were the opportunity to access climate finance; the desire to be recognised as a leader in this policy area; and the ability to move forward on national priorities. In Ethiopia stakeholders talked about using the climate change agenda to pursue and even boost their national ambitions. This tied in with dominant national storylines about what LCRD means, including the possibility of using LCRD to achieve better outcomes than those attainable through the use of other approaches. Rwandan stakeholders considered low-carbon resilience (combining both aspects of the climate change agenda) to be part of a broader national drive for environmental sustainability; they also recognised the need to demonstrate international leadership in this area.

When moving towards implementation, incentives relating to existing relationships, knowledge and finance come into play. For example, in Bangladesh the evidence relating to the extent of loss and damage caused by extreme climate events created a strong knowledge incentive to invest in infrastructure. Key to the decisions that followed, however, were the pre-existing partnerships between multilateral development banks and government line departments. This allowed the government of Bangladesh to rationalise the decision to use PPCR money to fund projects that were a good fit with this existing pathway. The availability of co-finance made a further economic case for this decision.

In Ethiopia SREP investment decisions were guided principally by economic incentives, policy goals and available knowledge. The decision to invest in large-scale grid-based electricity aligned with the economic incentive of co-finance and the policy incentives of national development plans for a fast-growing grid and energy export. Policymakers were also aware that climate variability was already affecting existing energy sources such as hydropower, creating an additional knowledge incentive for diversification of technologies. The IFC's existing risk-sharing facility with the International Bank of Ethiopia also represented a knowledge incentive, prompting the IFC to use the same model as a means of encouraging SMEs to invest in renewables.

In summary, we can see that the LCRD approach is shaped by existing national incentives and relationships, and policy goals and new knowledge can also play a role in determining which decisions are taken forward. Identifying the incentives that initiate and then shape the agenda can help us to understand which parts of the approach are likely to progress successfully to implementation, and which parts, if important, may need to be further incentivised.

Incentivising short- and long-term LCRD approaches

Currently, incentives to climate change action often prioritise short-term gains over the longer-term agenda. This poses a challenge, as a fundamental characteristic of an LCRD approach is that it plays out over long time frames. National plans that include both low-carbon and resilience agendas often focus on 'low-hanging fruit' rather than on a more profound integration of these two agendas, i.e. the policy focus is on short-term outcomes. Greater integration is likely to leverage improved outcomes and be more effective in the long run, but it is also likely to require greater policy innovation and cross-institutional cooperation, and different finance channels. It is therefore supported by fewer incentives and subject to greater organisational inertia.

Although programmes such as the CIFs seek to support 'transformational change' and national strategies depend on the availability of longer-term finance, some incentives continue to prioritise short-term outcomes. For example, the CIFs use co-financing (securing investments from third parties) as a measure of performance. This can discourage decision makers from selecting projects that are unlikely to attract co-finance but which may be more transformative in the long-term.

Similarly, of the two approaches to climate finance outlined in Chapter 6, the single channel modality is best suited to addressing short-term, project-specific priorities. In contrast, the multi-channel modality is better suited to long-term, programmatic and cross-sectoral approaches, and therefore to LCRD. Additional incentives for the latter are needed given the more far-reaching and therefore challenging nature of the approach.

Finally, pre-existing policies, relationships and processes can effectively incentivise short-term approaches by providing a comparatively easy path to implementation. For example, part of the motivation behind the selection of PPCR projects in Bangladesh was that the required administrative relationships and funding arrangements were already in place. It is also the case that existing structures can disincentivise change. For example, we saw that realising co-benefits between climate change and development agendas represented a particular challenge within national planning, as it meant disrupting established ideas, approaches and arrangements.

In summary, then, to support an LCRD agenda incentives need to be constructed around longer-term outcomes and synergies, and to counteract the disincentives inherent in existing relationships and pathways, which favour pre-existing approaches.

Aligning incentives across levels

Incentives work across levels – international, national and local – to influence decisions, and the extent to which such incentives are aligned determines whether they support or inhibit a particular course of action. One example of alignment across levels is the instance from Chapter 3 mentioned previously, where initial incentives for countries to engage with the LCRD agenda include the availability of climate finance and recognition as a policy leader, as well as the opportunity to address national planning priorities. The first two of these incentives operate at international level, the third at national level.

Chapter 6 provides a further example. High-level policy incentives related to energy access and social development established by the governments of Bangladesh and Nepal prompted national agencies – the central bank and IDCOL in Bangladesh and the AEPC in Nepal – to create economic incentives to invest in decentralised energy projects. These took the form of low-cost loans and institutional grants, and in turn they incentivised intermediaries such as commercial banks, microfinance institutions and suppliers to provide finance for the purchase of renewable energy products. At the level of the end user, the incentives to buy these products were socioeconomic, taking the form of co-benefits such as improved health and educational outcomes.

In this way, incentives can be aligned to support action and engagement across scales. However, this is not always the case – incentives at different levels may work against each other. For example, incentives at national level may encourage investments requiring long-term, sustainable finance, but at international level incentives are often still structured around short-term, political time frames, including finance with only a short- to medium-term investment horizon.

Structuring incentives to reach the poor

Reaching the poorest communities is fundamental to an equitable and inclusive LCRD agenda. Our findings suggest that the needs of the ultra-poor are not adequately served by current financing models, however. For example, the requirements of the Bangladesh Central Bank's Solar Home System and Solar Irrigation Pump schemes often cannot be met by the ultra-poor: they are unable to cover upfront costs or provide collateral for loans, and the available repayment periods and interest rates make loans unaffordable. Currently, there are no specific measures in place that provide for this group. We argue that incentives need to be specially structured so that LCRD policies deliver benefits for the poor and ultra-poor.

As we discussed earlier, having the appropriate actors with the right organisational capacities and mandates in place is important, and incentives are needed to structure the engagement of such actors. Higher-level policy and regulatory incentives play an important role in establishing further incentives such as policies, targets and fiscal measures to bring in the right actors, including financial intermediaries, microfinance institutions and SMEs, and end users. Regulatory incentives can help to reduce the risk of investment for the private sector. Economic incentives based on concessional financing are crucial in encouraging financial intermediaries such as private companies, commercial banks and microfinance institutions to invest. Microfinance institutions and NGOs are well placed to work with low-income communities in the earlier stages of market development. As well as an extensive reach, their social development orientation can incentivise them to make higher-risk investments in poorer communities, where commercial lenders' profit orientation may act as a disincentive. However, transaction fees can be high and mechanisms are needed to ensure that the finance they offer is affordable, and longer-term options may be needed.

Financial products also need to be tailored for the very poor, if end users are to be incentivised to use them. This may mean initially providing grants to build capacity in new markets, before moving on to a mixture of grants and concessional loans. Finance offered to lower-income customers should be appropriate to their particular needs; usually this means long-term, flexible finance with affordable repayment terms and limited security requirements. However, some very low-income customers are likely always to need some kind of grant support. Social protection systems may provide an appropriate way of ensuring that they are covered. Renewable energy products can also be adapted so that poorer people still have access to them, for example by reducing the size of a system and thereby its cost, while still meeting household energy needs.

To summarise, incentives for renewable energy access can be structured to reach the very poorest. This involves high-level policy incentives; recruiting actors with the right capacities and incentives to reach target groups; and tailoring the products and services so that they are attractive to end users.

Cross-cutting lessons for national LCRD policymaking

Based on the findings we have presented, we can draw out a number of cross-cutting lessons for national planning and the LCRD agenda. These can help LDCs to address the significant and novel challenges of an LCRD approach throughout the policy cycle, from agenda setting to prioritisation and implementation. We have organised the various lessons according to the stages of the policy cycle, and within this according to the particular challenges that the LCRD agenda presents.

Agenda setting

Adapting to a new and cross-cutting approach

- A range of stakeholders need to be included in the agenda-setting phase to ensure that diverse perspectives are included and that there is widespread buy-in. This helps to gain and sustain support for a new and cross-cutting LCRD approach. Different actors will have different levels of convening authority, sense of ownership, political will and organisational contexts, as well as different incentives and resources, and different levels of influence. These criteria need to be balanced to ensure both that the agenda has sufficient momentum to move forward and that diverse perspectives are taken into account.
- Questions about what an LCRD approach means for society and national economies need to be considered as part of agenda setting. In order to make effective decisions about climate funding, governments

and international climate finance initiatives need a thorough understanding of the internal political economy of these decisions. As well as institutional structures, mandates and incentives, this understanding should extend to issues of transformational change, long-term aims and policy trade-offs.

Ensuring equity and inclusion

- Including a range of actors with diverse and alternative viewpoints in agenda setting ensures that all voices are heard and are bought into the process
- Concerns about equity and inclusion need to be explicit from the beginning, and concepts such as 'win-win' outcomes, synergies between agendas and trade-offs need to be considered in relation to the national context to define a local understanding and approach. This can be done by problematising what has and has not been included in LCRD policies and why, allowing those involved to examine the assumptions underpinning these policies. Assumptions made about the effectiveness of processes and outcomes need to be interrogated in relation to experiences at national and local level, to ensure that their validity is tested rather than assumed to translate across scales. Related to this, key questions about who benefits from a particular policy approach and who is excluded or otherwise disadvantaged by it also need to be asked.

Prioritisation

Adapting to a new and cross-cutting approach

- An assessment needs to be made to see if more transformative approaches should be prioritised, and if so how existing policy networks, discourses and priorities will help or hinder this. National LCRD policymaking is heavily influenced by pre-existing national priorities, policy networks, institutions and incentive structures. As a result the changes it produces tend to be gradual and moderate rather than truly transformative. If actors want to push for a more radical agenda then a solid evidence base needs to be developed to support it. In particular, the evidence needs to show that the benefits of adopting the agenda will outweigh the disruption and cost involved in establishing the new relationships, structures and delivery models it will require.
- Dominant storylines or narratives need to be identified early on. This will help to predict where policies will find natural support and where there may be barriers to implementation, as well as with ensuring that key narratives are not being excluded. Those discourses supported by resources and incentives tend to shape prioritisation and investment choices.
- Approaches with longer-term outcomes need specific incentives to ensure that they are feasible choices.

Lessons from the political economy of the LDCs 167

Ensuring equity and inclusion

- As in agenda setting, diverse perspectives should be included at the stage of prioritisation, with actors being given the resources they need to participate meaningfully.
- Marginalised issues also need to be critically addressed in the prioritisation stage. These should include consideration of who benefits from the approaches chosen and who is excluded.

Implementation

Adapting to a new and cross-cutting approach

- Incentives and disincentives to the implementation of LCRD approaches need to be identified so that actors can understand how approaches will move forward and recognise any potential blockages to implementation. Different types of incentives are needed for different actors and activities. In addition to policy incentives, these include regulatory, economic, knowledge and reputational incentives.
- Incentives need to be aligned across scales to ensure smooth implementation.

Reconfiguring relationships

- Discourse coalitions that might emerge around a particular approach (but which have not yet taken the form of a stable group of actors) can be supported to develop into policy networks to take forward particular approaches based on a shared understanding. Supporting the development of entirely new policy networks – where they or potential precursors such as discourse coalitions do not yet exist – can help to coordinate approaches. Policy networks can be key to moving an agenda forward, and to supporting learning and iterative planning.
- Developing policy networks between actors at different levels – international, national and local – can be important for developing discourses and aligning incentives that support implementation and maintain momentum for LCRD actions.

Ensuring equity and inclusion

- Specific incentives are needed at all levels, from international to local, to ensure that appropriate actors with sufficient capacity, organisational flexibility and other incentives are engaged in ensuring energy access for the poorest communities. This issue needs to be explicitly addressed to ensure that an LCRD agenda does not leave the poorest behind.

- Renewable energy products need to be tailored to the needs and resources of the very poor, in order to incentivise their long-term use.
- While high-level policy incentives and key actors can initiate and support an LCRD agenda, critical reflection is needed at all levels to ensure that measures are effective and equitable.

Conclusions

Taken together, the findings presented here show that internal political economy factors are key to how a national LCRD agenda is realised. UNFCCC – and indeed individual states – tend to portray countries as unitary actors with consistent and well-aligned discourses and incentives within and between scales, but as we have shown this is often not the case. This has significant implications for how the objectives of the Paris Agreement will be achieved, and whether countries' INDCs will meet national and international objectives for all sections of their populations, including the poorest and most vulnerable.

Scholarly debate initially framed climate change as an international issue and focused on inter-state dynamics and regime theory, before moving on to consider its sub- and transnational aspects. We contend that it now needs to move on again, to unpack intranational issues and give these serious consideration in debates over how to achieve climate change objectives, and over the role of the state in climate governance. State actions are vitally important to the achievement of a low-carbon resilient future, and the interactions within and between states, and between states and sub-, trans- and international actors contribute to the complex, messy picture of climate governance. Jordan et al. argue for the 'rediscovery of the state as a dynamic site and catalyst of governing' within climate change governance (2015: 978), and identify national processes as one of three dimensions of a new climate governance. This supports the argument we make in this book that processes of governance within the state need to be developed further in theorisations of climate action. However, in their analysis Jordan et al. focus on broad factors underlying the adoption of national policies and much of their argument is based on mitigation measures in developed countries. We have shown that analysis needs to look beyond this to issues across the policy cycle, and to the actors, discourses and incentives that underpin internal state action. It also needs to examine experiences in all countries, including LDCs.

Issues of climate change in the LDCs highlight the necessity of policy trade-offs and lend real urgency to the matter of poverty alleviation and energy access. Climate change also challenges many of the norms of public policy, due to the multiple sectors, actors and scales involved, its uncertain evidence base and its long-term nature. It is not surprising, then, that LCRD policy and finance are contentious, or that incentives are not consistently aligned so as to deliver particular outcomes. However, following the 2015 Paris Agreement and with money flowing into the GCF, the time has come to shift the focus of climate action and debate to the national arena: to governments and their agencies, and to non-state and subnational actors. The next decade will be a crucial time if the strategies discussed in this book are to be effective in providing access to clean energy to those currently without it while also protecting communities, infrastructure and economies from the impacts of climate change. Our research suggests that progress is being made but that there are likely to be challenges ahead. We need to prepare now to meet these challenges, to ensure that appropriate trade-offs are made, barriers are overcome and finance reaches those who need it most.

References

Ayers, J., Anderson, S. and Kaur, N. (2011) Negotiating climate resilience in Nepal. *IDS Bulletin*, 42: 70–79.

Börzel, T. A. (1998) Organizing Babylon: on the different conceptions of policy networks. *Public Administration*, 76: 253–273.

DFID (2009) *Political economy analysis: how to note.* DFID Practice Paper. London: DFID. Available at http://www.odi.org/sites/odi.org.uk/files/odi-assets/events-documents/3797.pdf (accessed 22 December 2015).

Jordan, A., Huitema, D., Hildén, M., van Asselt, H., Rayner, T., Schoenefeld, J., Tosun, J., Forster, J. and Boasson, E. (2015) Emergence of polycentric climate governance and its future prospects. *Nature Climate Change*, 5: 977–982.

Rai, N. (2013) *Climate Investment Funds: understanding the PPCR in Bangladesh and Nepal.* IIED Briefing Note. London: IIED. Available at http://pubs.iied.org/pdfs/17151IIED.pdf (accessed 19 March 2016).

Tanner, T. and Allouche, J. (2011) Towards a new political economy of climate change and development. *IDS Bulletin*, 42: 1–14.

Index

actor-centred 10
actors and networks 14, 154–156
adaptation 2, 4
Adaptation Fund 24–25, 68–69
Alternative Energy Promotion Centre (AEPC) 111, 113
alternative narrative 74–76
Annex 1 countries 1

Bangladesh Climate Change Resilience Fund (BCCRF) 32
Bangladesh Climate Change Strategic Action Plan (BCCSAP) 31
Bangladesh Climate Change Trust Fund (BCCTF) 32
Bangladesh LCRD plan 30–33
base of the pyramid 106, 107
budget code 39, 40

Cancun Adaptation Framework 23
Central Bank of Bangladesh 113
climate change 1
Climate Resilient Green Economy (CRGE) facility 34, 94, 95, 97
climate resilient infrastructure 74
constructive national planning tool 135
cooperatives 117

decentralised energy 108–109
delivery agents 115–117
delivery models 111–115
Development Bank of Ethiopia 110, 114
Development Bank of Rwanda 114
developmental impacts 70
discourse analysis approach 144
discourse coalitions 11
discourses 11, 14, 157–160
domestic political economy 9, 131

dominant narratives 74–76
Drivers of Change approach 132

Economic and Development Poverty Reduction Strategy (EDPRS) 37
energy access 107
energy finance 107

finance for poor 122–125
financial instruments 118–119, 126
financial intermediaries 115–117, 126
Fonds National de l'Environnement (FONERWA) 37, 90, 91, 94, 97, 111, 114

governance and political economy approach 134
grants 118
green banking 116
Green Climate Fund (GCF) 24–26
Growth and Transformation Plan (GTP) 34

incentives 10, 14, 77, 120–122, 146–148, 161–165
Infrastructure Development Company Limited (IDCOL) 110, 112
institutional and governance mapping 138–140
Intended Nationally Determined Contributions (INDC) 23

Knowledge 12
Kyoto Protocol 23

LCRD in Ethiopia 33–35
LCRD in Nepal 38–41
LCRD in Rwanda 35–38
least developed countries (LCD) 2

Index

Least Developed Countries Fund (LDCF) 23–25
Least Developed Countries Work Programme 23
leveraging finance 97
loans 119
Local Adaptation Plan of Action (LAPA) 39
Low Carbon Economic Development Strategy 39
low carbon resilient development (LCRD) 2, 4
low-income communities 107

Market Development for Renewable Energy and Energy Efficient Products Programme (MDREEEP) 110, 114
Marrakesh Accord 23
micro finance institutions 116–117
Ministry of Finance and Economic Development (MOFED) 34
Ministry of Natural Resources (MINIRENA) 37
Ministry of Science, Technology and Environment (MOSTE) 39
mitigation 2, 4
multi-channel modality 96–98

National Adaptation Plan (NAP) 23–24
National Adaptation Programme of Action (NAPA) 24
national development banks 113
national LCRD plans 26
National Strategy for Climate Change and Low Carbon Development (NSCCCLCD) 36
Nationally Appropriate Mitigation Action (NAMA) 23–24
new political economy 10

off-grid 110
one-stop shop model 112
outcome mapping 136–137

passive resisters 159
phased subsidy approach 112
Pilot Programme for Climate Resilience (PPCR) 27, 70–78
policy networks 11, 100–102, 157
political economy 9
political economy analysis tool 134
political economy approach/ analysis 2, 9
Politics of Development framework 133
post structuralism 11
power analysis tool 134
power influence grid 142–143
prospective and retrospective analysis 136
public private investment models 110

rational choice approach 10
Rwanda Environment Management Authority (REMA) 37

scaling up renewable energy programme 78–84
single-channel modality 95, 96
social innovation 74
Solar Home Systems 110, 112–114
Solar Irrigation Pumps 110, 112–114, 117
Special Climate Change Fund (SCCF) 24–25
special purpose agencies 111, 112
stakeholder analysis 141–144
storyline 11, 14, 66
Strategic Programme for Climate Resilience (SPCR) 71
structural approach 10

transformational change 70

United Nations Framework Convention on Climate Change (UNFCCC) and Conference of the Parties (COP) negotiations 22–23